A Guide to Pharmacovigilance Audits in the USA

Heather Murphy

A Guide to Pharmacovigilance Audits in the USA

Springer

Heather Murphy
Holliston, MA, USA

ISBN 978-3-031-73595-0 ISBN 978-3-031-73596-7 (eBook)
https://doi.org/10.1007/978-3-031-73596-7

This Springer imprint is published by the registered company Springer Nature Switzerland AG
The registered company address is: Gewerbestrasse 11, 6330 Cham, Switzerland

Preface

The need for Pharmacovigilance (PV) audits have evolved in line with PV activities over the past few decades. As FDA PV regulations and the need for vendor outsourcing increased so has the need to provide effective oversight of the activities performed by vendors on behalf of pharmaceutical/biotechnology companies. It has now become mainstream practice to outsource some or all of a company's PV activities due to internal resource constraints as well as a rapidly changing global PV environment where specific expertise is often needed.

In this book, I will try to share the information I acquired and utilized throughout my experience with conducting audits within the pharmacovigilance/biotechnology industry in the United States. I am hoping to provide guidance on how a PV auditor can effectively plan, execute, and summarize PV audit activities based on my PV audit experience. This book is not intended to be an official guide to conducting a PV audit. Rather, the intent is to share my experiences and provide information, suggestions, and tips I have learned along the way. The focus will be on PV audits conducted in the United States. It will not discuss additional global, regional, and country specific PV requirements as they are extensive and not covered within the scope of this book.

Several years ago, when I began receiving requests to conduct PV audits, there were no reference materials specific as to how to conduct PV audits in the United States. There were numerous resources available that discussed audits from a Quality perspective but none that outlined the nuances associated with PV audits. Based on the resources available at the time, I needed to utilize the existing Quality standard for audit conduct and integrate these concepts with the requirements outlined in PV regulations, industry standards and best practices as well as my PV background. I developed my own audit approach and tools that aided me in conducting PV audits in a manner that provided the information needed for sponsor companies to effectively assess contracted vendor performance. In addition, many companies were starting to utilize external resources to evaluate internal company PV departments as the scope and complexity of internal PV activities has become more complex over time.

Throughout the course of this book, I will attempt to discuss a brief history of Pharmacovigilance, provide an overview of the most common types of PV audits, and then outline the process for PV audit preparation, audit conduct, and audit reporting activities. Along the way, I will provide some tips auditors can use as they move through these various activities. I will then summarize all of these discussions with a Dos and Don'ts table of a successful PV audit and conclude by providing some commonly used audit templates I have utilized for my PV audits.

Hopefully, this book will provide some basic guidance for any existing or new PV auditors. Each PV auditor has their own audit style for which this book is not meant to replace but perhaps assist with providing some insight on what other PV auditors are doing on a day-to-day basis. Each auditor approaches a PV audit from their own unique perspective based on individual expertise, industry experience, and auditing experience. I have merely presented what I have found to be successful for me and hope that in sharing this information it will provide some structural framework for any PV auditors.

Holliston, MA, USA Heather Murphy

Contents

Abbreviations

AE	Adverse Event
APIC	Active Pharmaceutical Ingredient Committee
AR	Annual Report
BIMO	Bioresearch Monitoring Program
CAPA	Corrective and Preventative Action
CBER	Center for Biological Evaluation and Research
CDER	Center for Drug Evaluation and Research
CDRH	Center for Devices and Radiological Health
CFSAN	Center for Food Safety and Applied Nutrition
CIOMS	The Council for International Organizations of Medical Sciences
COVID-19	SARS-CoV-2 virus
CRO	Contract Research Organization
CVM	Center for Veterinary Medicine
DMP	Data Management Plan
DQSA	Drug Quality and Security Act
eCRF	Electronic Case Report Form
eDMS	Electronic Document Management System
eLMS	Electronic Learning Management System
ESG	Electronic Submission Gateway
ETASU	Elements to Assure Safe Use
eTMF	Electronic Trial Master File
EU	European Union
FAERs	FDA Adverse Event Reporting System
FDA	Food and Drug Administration
FDAAAA	Food and Drug Administration Amendments Act
FDASIA	Food and Drug Administration Safety and Innovation Act
FDCA	Federal Food and Cosmetic Act
GDPR	EU General Data Protection Regulation
GVP	Good Pharmacovigilance Practices
HCP	Healthcare providers
ICDRA	International Conference of Drug Regulatory Authorities

ICH	International Conference on Harmonization
ICSR	Individual Case Safety Report
IFPMA	International Federation of Pharmaceutical Manufacturers and Associations
KPI	Key Performance Indicator
MAA	Marketing Authorization Application
MAH	Marketing Authorization Holder
MDN	Message Delivered Notice
MedDRA	Medical Dictionary for Regulatory Activities
MMP	Medical Monitoring Plan
MR	Medical Review
MRP	Market Research Program
MSA	Master Services Agreement
OSI	Other Safety Information
PADER	Periodic Adverse Drug Experience Report
PAP	Patient Access Program
PBRER	Periodic Benefit Risk Evaluation Report
PQC	Product Quality Complaint
PSP	Patient Support Program
PSUR	Periodic Safety Update Report
PV	Pharmacovigilance
PVA	Pharmacovigilance Agreement
QC	Quality Control
QMS	Quality Management System
REMS	Risk Evaluation and Mitigation Strategy
SAE	Serious Adverse Event
SAP	Statistical Analysis Plan
SDEA	Safety Data Exchange Agreement
SLA	Service Level Agreement
SME	Subject Matter Expert
SOP	Standard Operating Procedure
SOW	Statement of Work
SP	Specialty Pharmacy
SRP	Safety Reporting Portal
SUSAR	Suspected Unexpected Serious Adverse Reaction
WHO	World Health Organization
WHODrug	International dictionary of medicines created by the WHO Program and managed by the Uppsala Monitoring Centre

Chapter 1
Introduction

1.1 Introduction

This chapter will provide a foundational overview of the origin of the definition of Pharmacovigilance (PV), a brief historical overview of evolution of PV, and some common activities conducted within PV.

1.2 Pharmacovigilance Defined

The etymological roots of Pharmacovigilance are based on the Greek "pharmakon" (drug/medicine) and the Latin "vigilaire" (keep awake).

There are various definitions of PV, but they all contain similar fundamental elements:

The World Health Organization (2022) defines Pharmacovigilance as the "science and activities related to the detection, assessment, understanding and prevention of adverse drug effects or any other possible drug-related problems."

According to Fornasier et al. (2018), the European Commission defines PV as the "Process and science of monitoring the safety of medicines and taking action to reduce the risks and increase the benefits of medicines."

A contemporary definition of PV is the "process of monitoring the safety of medicines after they have been approved for use, including detecting, assessing, understanding, and preventing adverse effects or any other drug-related problems" (Bio Mapas n.d.).

H. Murphy, *A Guide to Pharmacovigilance Audits in the USA*,
https://doi.org/10.1007/978-3-031-73596-7_1

1.3 Goals of PV

The goals of PV are rather broad and wide reaching. A summary of some of the main goals are listed below.

- Monitor all safety information to identify previously unknown drug interactions and adverse effects
- Improve overall patient care, public health, and safety in relation to the use of medicines
- Protect the public and prevent risks through analysis of product effectiveness versus potential patient harm risk assessment
- Provide additional knowledge on the substance's effects where this information can be used to benefit patient care
- Promote an improved understanding of Pharmacovigilance through public education

As a result of the media focus on product withdrawals over the past several decades, PV activities have become more visible with the discussion focused on PV system activities and effectiveness, including the need for global harmonization of PV requirements. In the 1980s, the European Commission spearheaded the development of single market for pharmaceuticals. There was also discussion between Europe, Japan, and the United States on the feasibility of such harmonization. In 1989, during a WHO Drug Conference of Drug Regulatory Authorities (ICDRA), the plans for harmonization were initiated. Subsequent to this meeting, various regulatory authorities approached the International Federation of Pharmaceutical Manufacturers and Associations (IFPMA) to discuss a joint initiative and the International Conference on Harmonization (ICH) was conceived. In April 1990, the first ICH meeting was held in Brussels, Belgium, where regulatory and industry representatives of Europe, Japan, and the United States met to discuss the ICH mandate. This gave rise to what the pharmaceutical/biotechnology industry now refers to as the ICH standards and guidelines.

Although the focus of this book is specifically on the conduct of USA PV audits, it is notable that the FDA has adopted many ICH standards for PV so both cannot be considered separate and distinct in terms of PV activities.

1.4 Common PV Activities

Pharmacovigilance activities are multifaceted and vary depending on product requirements as well as regional and global regulatory requirements. Internally, each organization may manage these activities internally, outsource to an external vendor, or utilize a hybrid approach. In addition, depending on the size of the organization, PV activities are most commonly managed by a separate PV department, but it is not unusual to see some PV activities managed through other departments (e.g., FDA ICSR reporting managed by Regulatory). The primary PV activities discussed below may vary in title and responsibilities based on the organization. The examples below are used for illustrative purposes only based on my prior PV experience.

Some of the most common (but not all inclusive) PV activities in the United States include:

1.4.1 PV Operations

PV Operations is the foundation of any PV department and includes activities associated with the following:

- Adverse event (AE)/product quality complaint (PQC)/other safety information (OSI) identification, management, and submission (also referred to as independent case safety report (ICSR) management)
- PV quality and compliance
- PV safety system (databases) development, management, and maintenance
- PV audit/inspection readiness and support
- PV document development (SOPs, work instructions, forms, PVAs/SDEAs)
- Organizational PV education

The focus of PV Operations is not simply based on how each activity performed independently but also on how all PV department activities (including those departments outside of PV that impact PV) function together as part of the overall PV System. It establishes processes and manages how all safety data (pre- and postmarketing) are identified, collected, documented, and transmitted to the sponsor/business partner for analysis and regulatory reporting (if applicable). PV Operations also performs periodic analysis of the PV system to ensure the independent activities are functioning as expected, determine gaps, and formulate corrections to these gaps.

1.4.2 PV Science

PV Science activities focus on the analysis of safety data associated with a specific product. This includes cumulative periodic data analysis and the representation and analysis of this data within aggregate reports. PV Science participates (in some instances performs) in the product specific literature search strategy development and management. This PV function works collaboratively with PV Operations and PV Medical Review and plays an important role in signal detection activities. The PV Scientist has a multifaceted role in that this individual needs to have an in-depth knowledge of the product they are working on and may serve as the product point of contact on safety management teams. These individuals collaborate with the primary PV physician for the product on periodic and aggregate safety data analysis and safety reporting development strategy. They also participate in the development of regulatory inquiry responses when needed.

1.4.3 PV Medical Review

PV Medical Reviewers are responsible for the ongoing ICSR and an aggregate medical review of a product's safety data. This individual is responsible for the analyses and interpretation of PV data in relation to known product safety information to determine if certain adverse events pose a safety risk to patients. PV medical works collaboratively with PV Science and PV Operations to ensure the safety data are as comprehensive as possible for medical analysis.

Medical review of ongoing and aggregate safety data is vitally important within any PV department to ensure patient safety through the identification of safety concerns to formulate risk management strategies if needed. Causality assessment is also performed to assess the temporal relationship between an adverse event and the product. This includes a comprehensive evaluation of the patient's medical history, concomitant medications, and the known properties of the product. Such activities are complex due to multiple contributing factors. The causality assessment to a specific product will determine regulatory reportability for some products. Medical reviewers are actively involved in the interpretation and analysis of safety data output as part of signal detection activities.

1.4.4 PV Signal Detection

PV Signal Detection is a collaborative activity within PV requiring various types of expertise. PV Operations typically provides the data listings used for analysis to PV Science. PV Science prepares the data by consolidating data from various listings into data tables. PV Science also conducts the first review and analysis of the safety data and provides preliminary findings associated with potential safety signals to PV Medical Review. PV Medical Review will conduct an analysis of the data in relation to the known product information and determine if there are any identified potential patient safety risks or signals. Signal detection is usually performed on a periodic basis to identify trends in the safety information received. According to the World Health Organization (WHO)-Uppsala Monitoring Center (UMC) (2024), a safety signal is information associated with a product known or new side effects and is typically generated from more than a single report of a suspected side effect.

> The primary goal of signal detection is to detect previously unknown causal associations between medicines and unexpected events using various statistical and analytical methods (CIOMS Working Group 2010).

Common data sources for signal detection in the United States include data housed in safety databases (including AEs/PQCs identified in the scientific literature), the FDA Adverse Event Reporting System (FAERS), and the FDA Sentinel system.

References

Bio Mapas (n.d.) A History of Pharmacovigilance. https://www.biomapas.com/history-of-pharmacovigilance/

Fornasier G, Francescon S, Leone R, Baldo P (2018) An historical overview over Pharmacovigilance, International Journal of Clinical Pharmacy 40:744–747. https://doi.org/10.1007/s11096-018-0657-1

Report of CIOMS Working Group VIII, Geneva 2010 Practical Aspects of Signal Detection in Pharmacovigilance

World Health Organization (2022) The Importance of Pharmacovigilance: Safety Monitoring of Medicinal Products

World Health Organization-Uppsala Monitoring Center (2024). https://who-umc.org/signal-work/what-is-a-signal

Chapter 2
The History of Pharmacovigilance

The following is a brief overview of some of the major historical events that led to the need for Pharmacovigilance and forever changed patient safety oversight in the pharmaceutical industry.

Pharmacovigilance began in 1898, following the death of a young girl from Northern England who was administered a chloroform anesthetic before the surgical removal of an infected toenail. Some speculated that the chloroform anesthetic might have caused a cardiac arrhythmia or pulmonary aspiration that led to her death, but the actual cause of death was not determined Fornasier et al. (2018). Subsequently, on June 30, 1906, the United States Federal Food and Drug Act was formed and it was mandated that drugs must be pure and free of any adulteration.

During the months of September–October 1937, the first major public case of drug-related deaths in the United States occurred. A drug called Sulfanilamide, used to treat streptococcal infections at that time, was responsible for the deaths of 107 people. In early 1937, Sulfanilamide was used in a powder and tablet form. Due to increased demand for a liquid form of the medication, the manufacturer, S.E. Massengill Company, decided to develop a liquid form of Sulfanilamide. The company's chief chemist determined that sulfanilamide would dissolve in diethylene glycol; however, there was no requirement at that time to test for human toxicity. The mixture was tested for flavor, appearance, and fragrance but not for toxicity. Since human toxicity testing (safety studies) were not required, the company proceeded with compounding the elixir and distributed 633 shipments across the country. Unknown to S.E. Massengill Co., diethylene glycol was typically used as an antifreeze and was deadly poisonous to humans. Following the distribution of the sulfanilamide elixir, reports of numerous deaths were made to the American Medical Association (AMA). The AMA requested samples of the elixir for analysis. The AMA lab identified ethylene glycol and promptly issued newspaper and radio warnings that the sulfanilamide elixir was toxic and deadly. Given the absence of pharmaceutical safety requirements at the time, there was no federal requirement for the elixir form of Sulfanilamide to be pulled from the market (Ballentine, 1981)).

H. Murphy, *A Guide to Pharmacovigilance Audits in the USA*,
https://doi.org/10.1007/978-3-031-73596-7_2

As a result of these sulfanilamide elixir-related deaths, the Federal Food, Drug, and Cosmetic Act, which increased the FDA's authority to regulate drugs, was passed in 1938. The act required that new drugs must show they are safe prior to being sold (FDA, n.d.).

In 1951, the US Congress passed the Durham-Humphrey Amendment, which defined the kinds of drugs that could not be used safely without medical supervision. The amendment limited the sale of these drugs to prescriptions only by medical professionals.

In 1952, following a nationwide investigation by the FDA, which revealed that chloramphenicol, an antibiotic, caused nearly 180 cases of often-deadly blood diseases, the FDA engaged the American Society of Hospital Pharmacists, the American Association of Medical Record Librarians, and, later, the American Medical Association in a voluntary program of drug reaction reporting.

In 1957, thalidomide first entered the German market as an over-the-counter sleeping remedy. By 1960, thalidomide was marketed in 46 countries, and around this time, an Australian obstetrician, Dr. William McBride, began to prescribe the medication off-label for pregnancy-induced morning sickness. This was not an unusual practice at that time. In 1961, Dr. McBride observed an association between the mothers he prescribed this medication to for morning sickness and deformities in the babies he delivered. It was determined that the drug interfered with the babies' normal development, causing many of them to be born with phocomelia, resulting in shortened, absent, or flipperlike limbs. As a result of a report of 161 affected babies in a German newspaper, the makers of the drug stopped its distribution within Germany. Other countries followed suit, and by March of 1962, the drug was banned in most countries where it was previously sold. The US media reported on how Dr. Frances Kelsey, an FDA medical officer, helped prevent approval and marketing of thalidomide in the United States. These reports garnered public support for stronger drug laws in subsequent years. As a result, the US Congress passed the Kefauver-Harris Drug Amendments, which required drug makers to prove their drug worked before the FDA could approve for sale.

In 1970, the FDA required the first medicinal package insert where all medications must contain information about potential benefits and risks.

In 1982, the FDA issued Tamper-resistant Packaging Regulations to prevent poisonings following deaths associated with potassium cyanide–laced Tylenol in the early 1980s, which resulted in the deaths of seven people in the Chicago, Illinois area.

In 1983, the US Congress passed the Federal Anti-Tampering Act, making it a crime to tamper with packaged consumer products.

In 1993, the FDA launched MedWatch, a voluntary system designed to collect reports from health professionals on problems with drugs and other medical products.

In 1998, the FDA introduced the Adverse Event Reporting System (AERS), a computerized database designed to store and study safety reports on already marketed drugs.

As per Sibbald (2004), Merck & Co. announced on 30 September 2004 a voluntary worldwide withdrawal of rofecoxib (Vioxx) after a study showed patients taking the drug on a long-term basis face twice the risk of a heart attack compared with patients receiving placebo. It was later estimated that the drug had caused more than 27,000 heart attacks and sudden cardiac deaths. It led to increased scrutiny of other similar class drugs (COX-2 inhibitors) and led to changes in the way clinical trials were conducted and in the way the FDA evaluated the safety of drugs.

In 2012, the Food and Drug Administration Safety and Innovation Act (FDASIA) expanded its authorities to collect user fees from industry to fund reviews of innovator drugs, medical devices, generic drugs, and biosimilar biological products following an outbreak of fungal meningitis the same year, which was linked to the contaminated compounded drug product, which resulted in the loss of 64 lives and caused more than 751 illnesses.

In 2013, the Congress enacted the Drug Quality and Security Act (DQSA), which ensured greater regulatory oversight of facilities creating compounded drugs. It outlined the steps for an electronic and interoperable system to identify and trace certain prescription drugs throughout the United States.

References

Ballantine (1981) Taste of Raspberries, Taste of Death: The 1937 Elixir Sulfanilamide Incident" www.fda.gov

Food and Drug Administration (n.d.) A History of the FDA and Drug Regulation in the United States https//www.fda.gov/centennial/history/history.html

Fornasier G, Francescon S, Leone R, Baldo P (2018) An historical overview over Pharmacovigilance, International Journal of Clinical Pharmacy 40:744–747. https://doi.org/10.1007/s11096-018-0657-1

Sibbald B. (2004) Rofecoxib (Vioxx) Voluntarily Withdrawn From Market, CMAJ 26;171(9):1027–8

Chapter 3
The Need for PV Audits

The need for PV audits has evolved over the past few decades and is now a critical tool used by pharmaceutical/biotechnology companies to ensure compliance with regulatory requirements and contractual obligations with business partners and vendors and to prepare for pending regulatory inspections.

PV audits are considered an independent, systematic, documented, and disciplined process for obtaining evidence and objectively evaluating the evidence to determine the extent to which the audit criteria are fulfilled (EMA 2015). PV audits are driven by the provision of objective evidence to verify compliance with regulations, contractual requirements, and existing procedures.

Audits are intended to be a collaborative exercise by which the sponsor, auditee, and auditor can determine compliance against the established criteria based on the audit plan and audit agenda. It should provide a nonthreatening environment where issues can be discussed, education provided to the auditee, and remediation for the observations provided by the auditee for the benefit of the business. Audits are not meant to be a format for the airing of grievances or to take the place of daily sponsor oversight. Audits are not inspections as inspections are conducted by regulatory authorities who assess the compliance requirements where observations can lead to potential regulatory and legal penalties (APIC 2016).

3.1 Internal PV Audits

PV audits in the United States are typically performed to assess the status of an internal PV department's activities, as part of a sponsor vendor qualification program and as part of an ongoing sponsor vendor audit program, to determine business partner/license partner compliance with the PVA or SDEA and in preparation for an expected FDA inspection. These types of audits are discussed in more detail in Chap. 4.

H. Murphy, *A Guide to Pharmacovigilance Audits in the USA*,
https://doi.org/10.1007/978-3-031-73596-7_3

Each organization will usually determine its audit strategy based on a risk-based methodology. Commonly, this process is determined by the quality function within the organization based on regulatory requirements and industry standards. The annual audit plan outlines which internal department audits or external vendor or business/license partners (external) will be audited in that year. The higher the risk of the vendor or business/license partner's noncompliance with the organization, the more frequent the audit. High-risk vendors can be audited annually, while lower-risk vendors may be audited once every 3–5 years. Of course, the frequency may change if issues arise that pose a compliance risk to the sponsor organization.

The PV audit schedule is commonly integrated into the overall organizational annual audit plan, as quality is ultimately responsible for developing and executing the company's audit program. PV should have input into which PV vendors or business/license partners are audited and the frequency of such audits. PV is an important function of every pharmaceutical/biotechnology company, so the frequency of these audits may be different than audits in other parts of an organization.

3.2 Inspection Readiness Activities/Regulatory Requirements

Inspection Readiness

PV audits are a valuable tool to utilize as part of inspection readiness activities. Since the sponsor company is responsible for ensuring that PV activities performed internally or conducted by a contracted vendor meet the requirements as outlined in based on regulatory requirements, contracts, PVAs, SDEAs and internal procedures, these audits can provide a focused evaluation of specific PV activities to identifies strengths and weaknesses in PV system performance.

PV inspection readiness activities primarily focus on the typical PV activities, such as ICSR management, medical review, aggregate development, signal detection, and regulatory submissions; however, there are multifunctional internal stakeholders whose activities can impact PV compliance. The postmarketing PV environment has a broader cross-functional impact in which PV activities can be affected/impacted by several internal organizational functions, such as (but not limited to) marketing, commercial, medical communications, quality, medical affairs, and legal and regulatory. It is important to identify, evaluate, and analyze these stakeholders' activities in relation to the overall PV system.

Regulatory Requirements

PV audits ensure that activities meet requirements as outlined in the FDA/ICH regulatory requirements. Although the scope of this book is limited to the conduct of PV Audits in the United States, it is important to note that the FDA regulations should be considered in relation to other global regulations for products with multiple country marketing authorizations, cross-reporting requirements, etc.

FDA pre- and post-marketing inspections are based on specific compliance programs, which are used as the basis for FDA inspections to assess compliance with the Federal Food, Drug and Cosmetic Act and other laws administered by the FDA.

3.3 FDA Compliance Program

3.3.1 Compliance Programs Overview

The FDA compliance programs are organized based on various product type categories, including the following:

Biologics are monitored through the FDA Center for Biologics Evaluation and Research (CBER). This includes biological products such as blood, vaccines, tissues, allergenics, and gene and cellular therapies (FDA 2024a).

Bioresearch Monitoring is managed through the FDA Bioresearch Monitoring Program (BIMO). This program is unique and comprehensive, as it monitors all of the conduct and reporting of preapproval and postmarketing FDA regulated activities, including those for new drugs, medical devices, food and color additives, veterinary products, and US tobacco products. This compliance program is not limited to the United States but also extends internationally. The scope of the BIMO program extends across the FDA's six centers (CBER, CDER, CDRH, CFSAN, CTP and CVM) (FDA 2024b).

Devices and radiological health activities are monitored by the FDA Center for Devices and Radiological Health (CDRH). The mission of the CDRH is to protect public health by ensuring that patients and providers have access to safe and effective medical devices and radiation-emitting products. CDRH actively educates health professionals, industries, and consumers to provide up-to-date information so they can make informed decisions (FDA 2024c).

Drugs are monitored through the FDA Center for Drug Evaluation and Research (CDER). CDER ensures that drugs are safe and effective before they are available to the general public. The scope of CDER includes prescription and over-the-counter (OTC) and generic drugs and includes products such as antiperspirants, fluoride toothpaste, dandruff shampoos, and sunscreens. The center is responsible for overseeing the research, development, and marketing of drugs (FDA 2024d).

Food and Cosmetics are monitored through the FDA Center for Food Safety and Applied Nutrition (CFSAN). This center protects and promotes public health and is responsible for modernizing methods to find, track, and eliminate harmful germs and other hazards; evaluating the safety of new ingredients for food and the safety of new color additives; strengthening manufacturing practices; ensuring food, cosmetics, and dietary supplements are appropriately labelled; fostering good nutrition and effective food safety practices; investigating causes of foodborne illness outbreaks; and targeting unsafe products (FDA 2024e).

Veterinary medicines are monitored through the FDA Center for Veterinary Medicine (CVM), which is responsible for protecting human and animal health. Some of the responsibilities of CVM include ensuring animal drugs are safe and effective, properly made, and adequately labeled and packaged; ensuring food-producing animals are safe for people to eat; and investigating the side effects and product quality problems that are reported for animal food, drugs, and devices (such as thermometers and pacemakers) once they are sold on the market (FDA 2024f).

3.3.2 *Types of FDA Inspections*

A sponsor's FDA inspection will be performed by the center that has responsibility for oversight of a specific product type. There are different types of inspections that may be performed, including preapproval, routine, follow-up, and for-cause inspections.

Preapproval Inspections
Preapproval inspections are conducted as part of the application review process prior to the approval of a new drug, device, or biologic. This inspection is conducted to ensure the validity of information provided in the sponsor's company's application. Such inspections look at data associated with manufacturing, clinical trial management, and patient safety. The result of this inspection will go into the overall evaluation for product approval.

Routine Inspections
Routine inspections, also known as surveillance inspections, are typically conducted every 2 years to monitor the manufacturing process and quality of FDA-regulated products on the market. The agency uses the inspection to evaluate whether a marketing authorization holder (MAH) is complying with postmarketing requirements for their product.

Follow-Up Inspections
Follow-up inspections, also known as compliance follow-up inspections, are conducted to ensure the MAH has remediated previous inspection observations in accordance with FDA requirements. These inspections can be conducted depending on the type of observations identified through a previous inspection where a 483 form or warning letter was issued.

For-Cause Inspections
For-cause inspections are conducted as the result of complaints associated with a facility's products, potential safety issues or to evaluate corrections to previous violations. For-cause inspections usually focus on the issue of concern, but the scope of the inspection can be expanded as needed.

References

Active Pharmaceutical Ingredients Committee (APIC) (2016) Auditing Guide chrome-extension://efaidnbmnnnibpcajpcglclefindmkaj/https://apic.cefic.org/wp-content/uploads/2021/10/APIC_CEFIC_AuditingGuideAugust2016.pdf

European Medicines Agency (EMA) (2015) Good Pharmacovigilance Practices (GVP) Module IV: GVP Audits

FDA (2024a) About CBER: https://www.fda.gov/about-fda/center-biologics-evaluation-and-research-cber/about-cber

FDA (2024b) Bioresearch Monitoring Program Information: https://www.fda.gov/inspections-compliance-enforcement-and-criminal-investigations/fda-bioresearch-monitoring-information/bioresearch-monitoring-p

FDA (2024c) Center for Devices and Radiological Health: https://www.fda.gov/about-fda/fda-organization/center-devices-and-radiological-health

FDA (2024d) Center for Drug Evaluation and Research | CDER: https://www.fda.gov/about-fda/fda-organization/center-drug-evaluation-and-research-cder

FDA (2024e) Center for Food Safety and Applied Nutrition (CFSAN) https://www.fda.gov/about-fda/fda-organization/center-food-safety-and-applied-nutrition-cfsan

FDA (2024f) Center for Veterinary Medicine https://www.fda.gov/about-fda/fda-organization/center-veterinary-medicine

Chapter 4
Types of PV Audits

4.1 Types of PV Audits

In this chapter, we will discuss the most commonly occurring types of PV audits in the United States as well as their importance within the overall PV system. In addition, I will also discuss the important elements that need to be of focus within each specific audit type. It is important to remember that no two PV audits should be approached the same way as each has specific requirements.

4.2 Preapproval Audits

Preapproval audits are Pharmacovigilance (PV) audits conducted in preparation for a Food and Drug Administration (FDA) preapproval inspection (PAI). FDA PAIs can be scheduled but are often unannounced meaning they can just "show up" at an organization's office location during the period by which the organization is awaiting an approval decision on a sponsor product marketing authorization application (MAA). The PAI is a component of the FDA's decision-making process whereby verification of the sponsor's application is verified and evaluated in accordance with FDA requirements. An important part of the PAI is the evaluation of the identification, documentation, and transmission of safety information to the sponsor along with the sponsor's management of this information as per the FDA 21 CFR Part 312.32: IND Safety Reporting (2023a). PAIs are conducted in alignment with the Food and Drug Administration Compliance Program: Chapter 46—New Drug Evaluation: Preapproval Inspections 7346.832 (2022a). PAI objectives include assessing the marketing applicant readiness for commercialization, conformance to the application, and data integrity and commitment to quality and pharmaceutical development.

H. Murphy, *A Guide to Pharmacovigilance Audits in the USA*,
https://doi.org/10.1007/978-3-031-73596-7_4

These audits can vary in scope depending on the needs of the sponsor organization. Some audits can be limited to the management of individual case safety report (ICSR) activities, but others can extend to additional PV activities including contract research organization (CRO) activities, regulatory submissions, aggregate data analysis, electronic trial master file (eTMF), and safety database validation and management. All or some of these activities can be outsourced to an external vendor. The assessment of PV activities in this context is based on contracts and various documents developed for specific clinical trials which contain drug safety information and processes. Such documents can include (but are not limited to) the protocol, data management plan (DMP), serious adverse event (SAE) reconciliation plan, safety management plan (SMP), statistical analysis plan (SAP), medical monitoring plan (MMP), CRO standard operating procedures (SOPs), and business/license partner Pharmacovigilance Agreement (PVA) or Safety Data Exchange Agreement (SDEA). In addition, an assessment of how this information was managed in relation to the procedures will be reviewed. For example, it is important to understand during a PV audit how a clinical site manages the identification, documentation (electronic Case Review Form (eCRF) and paper form), and transmission of a serious adverse event (SAE)/Product Quality Complaint (PQC) to the sponsor. Upon receipt by the sponsor, the case management and submission process will also be reviewed. So this is tracking case management from the source through to submission to the FDA (if applicable) or to business/license partners.

Vendor oversight activities including the evaluation and documentation of such activities will also be important. It is important to recognize that all activities performed either by the internal PV team or any associated vendors are potentially in scope for the preapproval audit as these are the topics that will be assessed during a PAI.

A PAI may not be needed if sufficient information is available to confirm information provided in the application and there are no outstanding issues. The determination as to whether PAI is needed is based on risk evaluation as determined by the FDA.

Auditor Tip

As a consultant, I am frequently asked to perform preapproval audits to ensure the PV function is prepared for an FDA inspection. Such exercises provide valuable information to the sponsor in determining potential gaps in PV activities and the documentation of such activities. It is important to remember that no company is perfect. Inadvertent mistakes do happen, and some clinical trials can extend over a period of years. Occasionally, SAEs may not get submitted to the sponsor from the clinical site on time or an SAE may be submitted late to the FDA. The important factor is to identify that these issues occurred, report to the sponsor, and investigate as to why these errors occurred. As an auditor, it is your responsibility to communicate such concerns to the sponsor, educate as to why remediation is important and provide assistance if needed.

4.3 Routine Audits

Routine PV audits are audits conducted within the scope of the sponsor organization's risk-based audit vendor program. The frequency and scope of routine PV audits vary depending on contracted activities. Such audits are typically PV system related based on contracted activities. High-risk vendors (e.g., providing full PV services) are commonly audited annually, while low-risk vendors (e.g., market research program (MRPs) vendors) may be audited once every three years to five years. The scope of these audits is normally based on contracted services and activities conducted since the qualification audit (for new vendors) or since the previous audit. The sponsor may also identify areas of concern for which the auditor is requested to focus on. More details on the conduct of the audit will be outlined in Chaps. 6 and 7.

4.3.1 PV Audit Reference

Routine PV audits are based on the elements outlined in FDA Compliance Program Chapter 53—Postmarketing Surveillance and Epidemiology: Human Drug and Therapeutic Biological Products: Postmarketing Adverse Drug Experience (PADE) Inspections Compliance Program for Human Drug and Therapeutic Biological Products (7353.001).

Individual case safety reports or (ICSRs) and periodic report data contained within the periodic adverse drug experience report or (PADERs), periodic adverse experience reports (PAERs), periodic safety update reports (PSURs), periodic benefit risk evaluation reports (PBRERs), and annual reports (ARs) are reviewed. The FDA will inspect any entities involved in the mandatory reporting of postmarketing safety information for approved and unapproved prescription and nonprescription drugs, as well as therapeutic biologics regulated by the CDER.

The FDA considers all manufacturers, suppliers, or distributors named on the label of an approved prescription drug marketed for human use will be evaluated in terms of the management of all incoming safety information to comply with the requirements of 21 CFR 314.80 c(1)(iii). For biologics, 21 CFR 600.80 will be followed where any entity whose name appears on the label of a licensed biological product as a manufacturer, shared manufacturer, joint manufacturer, supplier, distributor, or any other participant involved in divided manufacturing must evaluate all incoming safety information to comply with the requirements of 21 CFR 600.80 c(1)(iii). As a result, all of these entities are required to develop written procedures for the surveillance, receipt, evaluation, and reporting of postmarketed adverse experiences.

4.3.2 Routine Audit Elements

Below are some of the most common PV activities that will be evaluated during a US PV audit. This is by no means all inclusive as some companies may retain some activities internally and share activities between business/license partners or outsource activities to third-party vendors.

- Written procedures

Written procedures including the sponsor and vendors procedures pertaining to ICSR management, literature search activities, medical review and evaluation, signal management, aggregate reports, database system utilization and management, regulatory reporting, record retention, and data security are some of the procedures that may be in scope for such audits will be reviewed. Such procedures can include policies, SOPs, work instructions (WIs), and job aid guides and forms. The extent of the review will be dependent on the scope of the audit and whether specific activities are contracted with external vendors. The procedures will be reviewed against the PV activities performed to ensure alignment.

- ICSR Management

The management of ICSRs is assessed from the beginning of the process when the information is identified and reported, through to the transmission and receipt. This includes case processing steps such as intake (identification of the ICSR and initial assessment), management (data entry, MedDRA and WHODrug coding, narrative generation, QC, and medical review), and reporting (business/license partner or the FDA). These activities will be assessed against existing written procedures, contractual obligations, PVAs/SDEAs, and regulatory requirements. Data sources can be spontaneous or solicited. Spontaneous data sources include healthcare professionals (HCPs), consumers, family members, and caregivers to name a few. Solicited data sources include organized data collection systems such as patient assistance programs (PAPs), patient support programs (PSPs), or any active solicitation of information from patients/consumers.

The review of general surveillance activities including how the sponsor monitors for potential sources of safety information from the commercial marketing, postmarketing clinical investigation studies, scientific literature review, medical information requests, PAPs, PSPs, marketing programs, business/license partner, firm sponsored social media, legal cases, and product quality complaints can be evaluated. In addition, an assessment of whether the sponsor is surveilling both domestic and foreign sources (if in scope) will also be made. This is an important consideration because, although an audit may be considered to be US based, some companies can hold the marketing authorization holder (MAH) in other countries, or through business/license partners where specific safety information may need to be reported to the FDA as part of ICSR reporting requirements. An evaluation of how a valid ICSR is determined, the determination of reporting time frames, and the process for following up on these ICSRs for additional information will also be considered.

Combined AE/PQC identification, documentation, transmission, receipt, and management will be evaluated. The process for identifying a PQC within an AE report or AE within a PQC report will be reviewed. For combined AE/PQCs, there needs to be a documented process outlining how these events are identified, reported, and exchanged. Such procedures shall include provisions for review to determine whether the complaint represents a serious and unexpected adverse drug experience which is required to be reported to the FDA.

- Reporting Time frames

Reporting time frames are determined based on the product life cycle (clinical trial versus marketed product) and regulatory authority time frames. Clinical trial reporting time frames for suspected unexpected serious adverse reactions (SUSAR) are 7 calendar days (for life-threatening and fatal events) and 15 calendar days for all other SUSARs. Postmarketing reporting time frames in the United States follow a 15-calendar day reporting time frame for all expedited reports for New Drug Applications (NDAs) and Biologics License Applications (BLAs) where the evaluation of the date of the receipt and the submission date to the FDA will be reviewed along with the timely submission of aggregate reports and other non-expedited reports. If an ICSR or aggregate report is submitted late to the FDA, a justification of why this report was submitted late should be provided. Business/license partners may have specific timelines outlined in the PVA/SDEA by which ICSRs and aggregate are to be received.

- Literature Search Activities

The development and approval of the literature search strategy including the periodic review and approval of the search strategy will be reviewed. This is an activity many PV departments will outsource; so depending on whether the sponsor or vendor procedure is used, this will be part of the evaluation. The computer system/applications used to perform the search, including the frequency of searches, output format, review of the output for AEs/PQCs, transmission of the AEs/PQCs to the party performing safety database management, abstract/full-text article, and record management will be assessed.

- Aggregate Safety Reports

The frequency of the submission of aggregate reports is dependent upon whether the product is approved under an NDA or a BLA. Such reports require specific information for the product which describes the safety information received during a specific reporting interval. Typically, following product approval, the reporting interval is quarterly for the first three years and annually thereafter unless otherwise specified by the FDA. A determination of whether the aggregate report (PSUR,

PADER, PAER, PBRER, or AR) contains all the required elements including submission timelines and record retention will be evaluated.

- Vendor Oversight

Vendor oversight is an important part of the PV system that will be evaluated during PV audits. As was previously mentioned, outsourcing in PV occurs frequently, but the sponsor is ultimately accountable and responsible for ensuring the activities of the vendors are being performed in a compliant manner as contracted, outlined in various procedures, and as required by the FDA. Outsourced activities can vary and can include only one PV activity or may include all PV system activities. A major component of this assessment includes which procedures are to be followed (the sponsor or vendor) or whether there are additional documents outlining the roles and responsibilities associated with performing such activities. How the sponsor ensures the vendor is following the correct procedures will be reviewed. Vendor oversight is an ongoing process in PV, so evidence on how this oversight is being performed and evaluated will be reviewed. PV vendor audits are only one component of vendor oversight as oversight activities occur daily.

- Electronic Submissions

Electronic submissions will be assessed to determine if safety report submissions are in an electronic format that the FDA can process, review, and archive as needed. A review of the system-generated confirmation notices from either the electronic submission gateway (ESG) or the safety reporting portal (SRP) as well as the procedures for correcting and resubmitting any submissions for which the message delivered notice (MDN) indicated that the submission was not accepted will be reviewed. Also, the method and location of MDN messages will be considered.

- Record Retention

Record retention activities will be reviewed to ensure the sponsor/vendor has a procedure in place to address the retention period for safety records as required by the FDA.

Auditor Tip

The vast majority of the audits I have conducted have been "routine" PV audits, but it should be noted that even though these audits are considered routine, issues can arise during the audit that may need to be escalated to the sponsor. "Routine" in this context means that the vendor was categorized (at a moment in time) during the initial qualification audit or a subsequent routine audit. This does not mean there were or are no potential issues. Always be vigilant!

4.3.3 *For-Cause Audits*

For-cause PV audits are those audits conducted in response to a trigger or concern with the PV services being provided by a vendor or business partner. The trigger may be a noncompliance with regulatory requirements or contractual obligations which are considered to be an identified risk to the sponsor organization. Common reasons for these audits include specific complaints made to the FDA, concerns of stakeholders involved in the research process, significant issues identified during research activities, multiple deviations, or FDA-identified events of special interest.

As a result, the scope of these audits is usually specific, but the scope can be extended as needed to assess the issue that led to the for-cause audit. A typical agenda for a for-cause PV audit is usually narrow in scope, but the auditor has the ability to explore additional areas not previously outlined in the agenda to identify the area of concern and evaluate objective evidence related to the noncompliance. The time frame to notify the auditee of these audits is typically much shorter than a routine audit. In fact, many companies will have language already included in the vendor Master Service Agreement (MSA), business partner Pharmacovigilance Agreement (PVA), or Safety Data Exchange Agreement (SDEA) outlining the notification timeline for a for-cause audit. Routine audits may require the auditee notification of 30–90 days, while for-cause audits may only require auditee notification of 2–4 weeks (or less).

In my experience, for-cause PV audits are usually requested based on regulatory or contractual requirements not being met, which could impact patient safety. Most commonly, the PV issue surrounds the identification, exchange of information, regulatory reporting, and documentation of safety information by a vendor or a business partner.

Depending upon the outcome of the for-cause audit, there may be communications that will need to occur between the sponsor and the FDA. This could be in the form of newly identified ICSRs submitted to the FDA (that may not have been previously submitted) or additional communication with the FDA as needed.

Auditor Tip

The notification of a for-cause audit to the auditee should be made by the sponsor before the auditor has communication with the auditee. As the PV auditor, you are the objective participant in this process. Sponsors will identify the need for such an audit based on their risk evaluation and, as a result, are responsible for communicating their concerns to the applicable vendor or business partner. Such audits can be stressful for all parties, but before you have communication with the auditee, you should ensure this notification has been completed by the sponsor. You do not want to be trying to establish communications with a vendor or business partner without them understanding the urgency and importance of such an audit.

4.4 Vendor Qualification Audits

PV vendor qualification audits are audits performed by the sponsor as part of their vendor management program which is typically part of the organization's quality management system (QMS) in collaboration with the functional area subject matter experts (SMEs). Commonly, companies utilize a risk-based approach to determining which vendors are to undergo a qualification audit and the type of such audits. As a result, depending on established severity, audit types may consist of a questionnaire, a questionnaire with a review of vendor procedures or a questionnaire, vendor procedure review, and a full PV audit. The established risk categorization is based on the vendor's risk to the sponsor. Therefore, the risk will determine the type of audit to be conducted. For example, a vendor who will be performing full-case processing activities (e.g., triage, intake, narrative creation, QC, medical review, follow-up, and regulatory reporting) as well as literature search, aggregate report generation and regulatory submission, and signal detection activities would be considered a higher risk than a vendor who managed a small market research program. In essence, the number and type of activities being outsourced determine the supplier risk. Vendor qualification audits are guided by 21 CFR 312.52 Transfer of Obligations which regulates transfer of obligations between contract research organization and sponsors. For additional information, you can view the FDA Code of Regulations 21 CFR 211: Current Good Manufacturing Practice for Finished Pharmaceuticals and 21 CFR 820.22: Quality Audit and the International Conference on Harmonisaton of Technical Requirements for Registration of Pharmaceuticals for Human Use (ICH): ICH Harmonised Tripartite Guideline Q10: Pharmaceutical Quality System.

Vendor qualification audits are needed to ensure that a potential vendor has the appropriate infrastructure, skills, and experience to comply with regulatory and contractual obligations. The selection of an appropriately qualified PV vendor can reduce potential risk to the sponsor organization. It is important to recognize that qualification audits have limitations. During these audits, you are auditing the vendor infrastructure and existing internal procedures. You are normally not auditing existing sponsor activities as these audits are commonly performed before the vendor conducts any sponsor activities. Therefore, these audits are not an accurate representation of how the vendor would perform sponsor activities. It would only provide information as to how they could conduct sponsor activities. Audit questionnaires do not provide the level of detail needed to understand how a vendor would perform a sponsor's contracted activities. As a result, I most commonly see a combination of an audit questionnaire combined with a full qualification audit where vendor procedures are reviewed, and the vendor can discuss how they would perform specific PV activities.

Auditor Tip

PV qualification audits need to be customized to the services the vendor will be contracted to provide. The audit approach for a vendor potentially providing full services PV activities is quite different in scope than a vendor providing one or two limited services. As an auditor, you must consider the purpose and scope of these audits and modify your strategy, audit plan, and agenda accordingly. It is also important to discuss with the sponsor whether the scope of services may expand in the short term as this is valuable information to have for future activities.

4.5 Business/License Partner Audits

Business/license partner audits are sponsor audits of business or license partners who are involved in the identification, exchange, management, reporting, and communication of safety information. These audits fall within the overall annual audit plan and can be conducted on a routine basis (based on the identified risk to the sponsor organization) or as part of a for-cause audit. These audits are important as they allow the sponsor company, which can be the marketing authorization holder (MAH) for a specific product globally or in specific territories, or the business partner may be the MAH globally or in specific territories, to evaluate the partner's compliance with the business/license partner agreement and the PVA/SDEA. The sponsor's quality department in collaboration with PV will determine the frequency and need for such audits. The requirements for these audits are usually outlined in the sponsor and business partner PVA or SDEA. These documents outline the roles and responsibilities of the sponsor and business/license partner for activities such as (but not limited to) territorial responsibilities (countries where the MAH is held for each party), contact persons, ownership of the global pharmacovigilance system, the need for a local pharmacovigilance system, training responsibilities, signal detection, management of the reference safety information RSI, periodicity for audits for both parties, communication during regular inspections, and the requirements for compliance activities between both parties.

These audits can be complex as you will need to identify which party (sponsor or business partner) holds the MAH for the product in scope as well as which party holds the primary responsibility for each PV activity outlined in the PVA or SDEA. This is critical to understanding the elements to be audited. The PVA/SDEA may not be comprehensive or timely in that it would have been developed at a specific moment in time and may not have specific current activities. It is also important to understand whether clinical trials for that product are in scope for the audit as this is not always evident in many PVA/SDEAs I have reviewed over the years. Many sponsors write these documents to encompass any conceivable scenario that may occur with the business/license partner. They do not necessarily reflect which partner is responsible for each activity. This is not optimal for the auditor as you cannot discern true ownership over specific activities over time.

In the United States, PVAs/SDEAs are often used interchangeably, but this is not the case in other regions of the world. In my industry experience, I have known PVAs to be extensive documents outlining the roles and responsibilities for each contracted activity, whereas SDEAs are usually shorter documents that describe ICSR and aggregate reporting timelines and responsibilities. Below, I have listed some similarities and differences between PVAs and SDEAs I have most commonly seen. This list is not exhaustive and is meant to be illustrative.

Similarities and differences between PVAs and SDEAs in the United States:

PVA	SDEA
Contact information	Contact Information
Cooperation between parties in terms of audits, regulatory inspections, issue clarifications, safety cases, records, safety personnel, and use of third-party vendors	Cooperation between parties in terms of audits, regulatory inspections, issue clarifications, safety cases, records, safety personnel, and use of third-party vendors
Confidentiality	Confidentiality
Data exchange timelines	Data exchange timelines
Definitions and abbreviations	Definitions and abbreviations
Description and purpose of the document	Description and purpose of the document
ICSR handling and medical review and assessment	ICSR handling and medical review and assessment
Literature review activities and responsibilities	Literature review activities and responsibilities
Periodic reporting responsibilities	Periodic reporting responsibilities
Primary safety database owner and whether a local safety database would be needed	Primary safety database owner and whether a local safety database would be needed
Process for reconciliation activities; frequency, sources, process, documentation, and archiving	Process for reconciliation activities, frequency, sources, process, documentation, and archiving
Reference safety information (RSI) development, management, and training	Reference safety information (RSI) development, management, and training
Responsibility for ICSR follow-up	Responsibility for ICSR follow-up
Audit and inspection communications	
Business continuity and disaster recovery	
Conflict management	
Management of regulatory authority inquiries and communication with the partner	
Noncompliance or breach of contract	
Pharmacovigilance System Master File (PSMF) management	
PV system management	
Qualified Person Responsible for Pharmacovigilance (QPPV) responsibilities	
Regulatory monitoring	
Risk management	
Safety governance	
Signal detection management and communication between parties	
Urgent safety communications management	

Auditor Tip

It is important to clearly establish whether all tasks within the scope of a business/license partner audit are outlined in the PVA/SDEA. Both of these documents, from a PV perspective, are intended to represent the day-to-day PV responsibilities for a product(s). If the PVA/SDEA has not been reviewed and updated over a period of several years, it is reasonable to expect that there have been operational changes that may not be reflected. Always discuss the scope of responsibilities with the sponsor during audit planning meeting as this will have a critical impact on your development of the audit plan, audit agenda, and preaudit document request list. It is better to put some time into this discovery during the planning portion of the audit rather than be surprised during the audit conduct. This will allow you the opportunity to effectively formulate your audit strategy prior to the audit.

4.6 Contract Research Organization (CRO) Audits

CRO PV audits are important to ensuring the outsourced PV services are compliant with contractual expectations as well as regulatory requirements. The frequency of these audits follow the risk-based assessment performed by the sponsor's quality function and are intended to ensure the activities performed meet sponsor expectations as outlined in various operational documents, standard operating procedures (SOPs), work instructions (WIs), and contracts.

At this time in our industry, it is very common for pharmaceutical/biotechnology companies to outsource all or some of its PV activities. The reasons for this practice are varied, but outsourcing can fill gaps in specialized skills, technology, or knowledge absent within the organization. In some instances, outsourcing can provide a cost-effective option to establish all PV activities internally at the sponsor company. The core competencies of CROs vary across the development life cycle. Some CROs are quite adept at clinical trial management, while some have expertise spanning clinical trials to postmarketing activities globally. Regardless of what stage of the product life cycle a company may be in, the CRO, like any vendor, needs to be audited as part of the sponsor's oversight responsibilities.

One of the complexities involved in conducting these PV audits surrounds whether the sponsor company delegated certain activities to the CRO, whether the sponsor requested the CRO follow the sponsor company procedures, or whether the sponsor agreed that the CRO could follow their procedures. Either way, the sponsor is responsible for ensuring that the selected procedures are followed and that these procedures are compliant with FDA expectations. Sponsors need to ensure CRO expectations are clearly communicated prior to project launch and regularly throughout the course of the engagement. CROs cannot perform contracted services to the specificity of each sponsor without knowing expectations.

Some CROs take the position that they are not subject to FDA inspection as only the sponsor will be inspected. This is not correct. As a consultant, I have seen many CROs that have been inspected for various reasons specific to contracted services. It is commonly thought by sponsors that once the CRO signs the contract with the sponsor, they are expected to comply with regulations associated with the activities they are providing to the sponsor and by extension can be inspected by the FDA.

Auditor Tip
You may encounter some sponsors who consider PV audits to be the only method of providing CRO vendor oversight. This position is fundamentally flawed as audits are performed periodically (every 1–5 years). Sponsors are expected to provide and show evidence of ongoing, daily vendor oversight. PV audits are an additional tool that provides valuable insight into a "point in time" assessment of compliance. They do not provide granular daily oversight. Ultimately, the sponsor is responsible for providing ongoing, measurable, and documented oversight of all contracted vendor activities.

As the auditor, you may encounter CROs that are inexperienced in terms of their responsibilities with the sponsor and the risk for FDA inspections. There is no doubt based on the various audits I have conducted that CROs are inspected regularly. This is evidenced by the regulatory inspection history provided as a routine document request for PV audits.

4.7 Specialty Pharmacy Audits

Specialty pharmacy (SP) audits are audits of pharmacies that have expertise in the fulfillment and dispensing of specific medications used to treat serious or rare health conditions. Some SPs also provide additional services including patient/consumer enrollment assistance, co-pay assistance, benefit verification, nurse, and pharmacist assistance, as well as the generation of various client reports related to product dispensing, status, and inventory activities. SP audits are performed to evaluate contracted activities as well as compliance with FDA regulatory requirements related to adverse event and product quality complaint (AE/PQC) management and the computer systems where this information is stored, managed, or transmitted. It is true that SPs are primarily regulated by the individual state Board of Pharmacies for which they practice, but SPs are obligated to follow the FDA regulations for contracted services as agreed to with the sponsor. For example, if the vendor does not follow 21 CFR Part 314.80: Postmarketing Reporting of Adverse Drug Experiences, 21 CFR 211.198: Compliant Files, and 21 CFR Part 11: Electronic Records, Electronic Signatures, then the sponsor is noncompliant with these regulations.

SPs have become more prevalent following the passage of the Orphan Drug Act in 1983. The Orphan Drug Act is a law passed by Congress in 1983 that provided incentives to pharmaceutical companies to develop drugs to treat rare diseases. According to the FDA (2024b), a rare disease is a disease or condition that affects less than 200,000 people in the United States. Companies can request an orphan drug designation from the FDA and, if approved, may be entitled to certain incentives. As a result, sponsors can dedicate more resources to research and development for diseases that impact a much smaller portion of the population. As outlined in IQVIA (2018), there are approximately 7000 rare diseases that affect 25 to 30 million people in the United States. Many of these diseases can be life-threatening, but many can now be treated and managed as chronic conditions.

Due to the increased utilization of SPs, sponsors are required to include them in their annual audit plan. SPs are required, like any other contracted vendor, to be compliant with applicable regulations associated with the identification, management, and reporting of safety information. This also includes the provision of associated procedures to assist with these activities.

Auditor Tip
Although the SP landscape has evolved over the past few decades, you may encounter some SPs that will take the position that they do not need to comply with any FDA regulations. This is not correct. In terms of the identification, documentation, and management of safety information, including the computer systems utilized, they are expected to comply with the applicable FDA regulations. If you encounter this resistance, you should educate the SP as to why they are required to follow specific regulations and inform the audit sponsor. The sponsor will need to address this issue with the vendor subsequent to the audit.

4.8 Distributor/Supplier/Warehouse Vendor Audits

Distributor/supplier/warehouse PV audits are audits of vendors who are contracted to provide inventory management/supply/distribution services for a sponsor company' product. As such, these vendors are expected to have procedures in place for the identification, management, and communication of AE/PQC information to the sponsor. Since the employees at these vendors do not have direct contact with healthcare professionals (HCPs) or patients, the chance of identifying AE/PQC information remains low but not negligible. The sponsor should have a safety data exchange agreement established with the distributor/supplier/warehouse vendor to ensure the roles and responsibilities and time frames associated with providing this safety information are clear. The objective of these audits is to ensure the vendor is complying with contractual and regulatory requirements. The frequency of these audits will also be determined by the sponsor organizations' quality audit program.

Auditor Tip
The auditor will need to assess the vendor infrastructure to ensure compliance with the identification, management, communication, and document retention of safety information in reference to the type of vendor and activities performed. In my experience, these vendors do not have safety databases as the volume of AE/PQCs is minimal. However, there must be a controlled and trackable method of documenting AE/PQCs. Some of these vendors are not familiar with the need for appropriate documentation associated with AE/PQC management. Patience is key as additional education may be needed.

4.9 Risk Evaluation and Mitigation Strategy (REMS) Audits

Risk evaluation and mitigation strategy (REMS) audits are unique to the United States and are a requirement by the FDA for products that need to have additional safety information provided or additional oversight of safety activities by the sponsor organization for products approved in the United States. The FDA REMS Compliance Program requirements are outlined in the Food and Drug Administration Compliance Program 7353.001c; Chapter 53—Post Marketing Surveillance and Epidemiology: Human Drug and Therapeutic Biological Products. The document outlines the expectations for the sponsor company prior to and during the marketing phase of a product.

The FDA will determine if a REMS for new product is needed as well as the elements to be included as part of the REMS. They will notify the drug manufacturer of the need for a REMS and request that the sponsor provide a proposed REMS for review. Once received, the FDA will review and, if acceptable, approve the REMS. The sponsor is then responsible for implementing the requirements of the REMS within specific time frames.

REMS is a relatively recent requirement. On September 27, 2007, the Food and Drug Administration Amendments Act (FDAAA) enacted public law 110-85 which authorized the FDA, under section 505-1 of the Federal Food and Cosmetic Act (FDCA), to require any entity submitting or holding certain drug and biologic applications to submit a proposed risk evaluation and mitigation strategy (REMS) as part of the application process.

There are various requirements to satisfy the need for the REMS which will be determined by the FDA. Such requirements can include the need for medication guides, communication plans, or elements to assure safe use (ETASU). Additional details about the implementation system are included in the supporting document which is a document provided by the firm that includes a thorough explanation of the rationale and supporting information for the content of the rims but is not part of the approved REMS.

Applicant holders for NDA and BLA are required to submit assessments according to intervals set forth in the approved REMS. Minimal time frames can be 18 months, 3 years, and 7 years from the date of the initial approval of the REMS. The Center for Drug Evaluation and Research (CDER) Office of Scientific Investigations leads enforcement efforts when firms do not comply with their REMS requirement.

Since each REMS is unique, the different elements outlined in the approved REMS requirements letter will be addressed during the inspection.

4.9.1 Medication Guide

The REMS for a specific drug may require a medication guide that sets forth the requirements for patient labeling for human prescription drug products including biological products as outlined in 21 CFR 208: Medication Guides for Prescription Drug Products. For a REMS, the medication guide is required to be available for distribution to patients at the time the drug is dispensed. During a REMS audit, confirmation that a copy of the medication guide distributed with the drug product as well as verification that the copy sent is identical to the copy submitted to the FDA will be reviewed. This review will also include documentation of the methods the sponsor will use or has used to distribute the medication guide, verification of the availability to the end user, documentation the sponsor utilizes to identify, report, and correct the inconsistencies in complying the distribution and dispensing requirements. In addition, documentation of the sponsor's activities related to the assessment of the healthcare providers and patients' understanding of the messages communicated in the medication guide will be assessed.

4.9.2 Communication Plan

In some instances, a REMS communication plan may be required. The communication plan is directed at healthcare providers and provides information on how the product risks are addressed by the REMS. Some of the components of the communication plan include the REMS Website, dear healthcare provider letters, letters addressed to healthcare providers through the professional organizations, REMS fact sheets, patient counseling tools, or scientific journal information. During the PV audit, the focus of the communication plan will include the date of product launch and the source for mailing lists used to distribute letters to target the audience, which can include physicians and pharmacists. Verification of this information against the REMS document including the mailing lists used to distribute letters to the target audience, the distribution dates and method of distribution for the communication plan tools, and the management of undeliverable emails and faxes or returned postal mail for each distribution. The development of a corrective action plan to address the undeliverable mailings will also be reviewed. Functional activity of the REMS Website including the go-live date, Website issues, and management will be considered.

Elements to Assure Safe Use (ETASU)

The FDA may determine that a medication guide and communication plan are not sufficient to mitigate potential safety risks for a specific product. In such an instance, an ETASU REMS may be required. It should be noted that not all products requiring a REMS will require all subparts of the ETASU program as outlined below. The sponsor's REMS requirements will be determined and communicated to them by the FDA. Below is a brief summary of the various ETASU sections.

ETASU A (Section 505-1(f)(3)(A))

This section pertains to the requirement that healthcare providers who prescribe the sponsor's drug or have specific training or experience are certified. The audit will include an assessment of the training and documentation of training for healthcare providers in compliance with the terms of the REMS, documentation of the firm's operational activities related to the implementation of ETASU A (e.g., HCP notification that they have been certified in the REMS program and management and maintenance of a validated, secure database containing certified healthcare provider information. How the sponsor manages those HCPs that are noncompliant with the terms of the REMS, documentation of HCP understanding of messages communicated by the REMS program (e.g., knowledge surveys), and the sponsor activities for conducting surveillance associated with the REMS program risks will also be considered.

ETASU B (Section 505-1(f)(3)(B))

This section requires that pharmacies, practitioners, or other healthcare settings that dispense the drug are specially certified or have the opportunity to be certified for any willing provider. The following topics will be reviewed during a PV audit: confirmation that pharmacies, HCPs, or other healthcare settings that dispense the drug is specially certified as required by the REMS program, evidence that the certification process is in accordance with the approved REMS, documentation of the sponsor's activities related pharmacy, HCP and healthcare setting certification and evidence that the computerized system used to manage and retain this information is validated and secure, process for and evidence of how the sponsor manages the pharmacy, HCP or healthcare setting noncompliance, documentation of pharmacy, HCP or healthcare setting knowledge of the REMS program through targeted knowledge surveys or site audits, and confirmation that the sponsor is performing ongoing surveillance activities to monitor the risks associated with the REMS.

ETASU C (Section 505-1(f)(3)(C))

Under this section, it is required that the sponsor drug be dispensed to the patient only via specific healthcare settings, such as hospitals. During the PV audit, the following areas will be reviewed: confirmation that the drug was shipped to certified facilities with requirements outlined in the REMS, evidence of documentation of a distributor/wholesaler process as required by the REMS, a process for and documentation of the sponsor's management of distributor/wholesaler noncompliance, documentation of the distributor/wholesaler compliance with the REMS program and evidence of the sponsor's ongoing surveillance of the risks associated with the REMS program.

ETASU D (Section 505-1(f)(3)(D))

In this section, the requirement includes that the sponsor drug is dispensed with evidence or documentation of other safe-use conditions, such as the completion of laboratory testing or the need to have an EpiPen in the event of an emergency. During the PV audit, the auditor will determine if the sponsor product was dispensed to patients only with confirmation and documentation of other safety use conditions as outlined in the REMS, evidence that the patients were informed of the potential product risks and benefits prior to use and evidence of educational materials provided and confirmation that the product was dispensed following a specified authorization process, process for and documentation of the sponsor management of the noncompliance by the pharmacy, HCP, patient, or healthcare setting, evidence that certified prescribers can submit completed forms documenting safe use conditions as specified, confirmation that certified pharmacies or healthcare facilities can verify documentation of safe use conditions, management and maintenance of a validated, secure database and a REMS Program Call Center (if applicable). In addition, evidence that the sponsor conducts ongoing surveillance activities relating to the REMS risks will be evaluated.

ETASU E (Section 505-1(f)(3)(E))
Under this section, each patient may be subject to individual monitoring. As a result, the PV auditor will evaluate that each patient receiving the product received the specified monitoring as outlined in the REMS, sponsor process for and documentation of the pharmacy, HCP, patient, or healthcare setting noncompliance, documentation of the sponsor's implementation activities (e.g., checklist, monitoring schedules, etc.), and ongoing sponsor surveillance activities related to the risks identified in the REMS.

ETASU F (Section 505-1(f)(3)(F))
This section discusses the requirements associated with patients receiving product to be enrolled in a registry. REMS registries may be either mandatory or voluntary depending on the requirements of the approved REMS program. A mandatory REMS registry requires that the patient is enrolled in the registry prior to receiving the product. There are various data points that may be collected and analyzed including clinical outcomes, laboratory and clinical data, safety information, compliance, and the effectiveness of various tools used to ensure patient compliance and outcomes. Voluntary REMS registries may be required for a REMS where long-term safety data is collected. Patients are not required to participate in such registries in order to receive the product.

The PV auditor will confirm whether patients are enrolled in a registry based on the approved REMS program requirements, verify the sponsor's management of a patient registry noncompliance (may be managed by a third-party vendor), confirm that all patient registry data is managed and retained within a validated and secure database, as well as confirm the sponsor's ongoing surveillance activities associated with the REMS program-identified risks.

REMS PV audits can be requested prior to product launch or as part of the ongoing REMS program periodic assessment. Prelaunch REMS audits assess whether all the REMS program requirements are in place prior to the patient receiving the product and are compliant with the REMS program requirements. It provides the sponsor with information to determine REMS program development progress and areas for additional improvement. Postlaunch (ongoing periodic assessment) REMS PV audits evaluate the implementation and compliance of the required REMS program activities. They provide the sponsor with an objective assessment as to whether the REMS program is being implemented as expected.

Auditor Tip It is important for the PV auditor to consider that the FDA REMS Compliance Program documents are an effective resource and tool to use to conduct the REMS audit, but it should not be the only tool. There are specific activities that must be in place as per the REMS program, but you must not forget basic PV audit principles including AE/PQC reporting and computer systems that manage, transmit, and store safety data. Due to the complexity of implementing and managing REMS programs, sponsors often use several vendors to assist with specific tasks. Each vendor should be audited separately to determine their compliance with contracted activities. It is helpful for the PV auditor to have a comprehensive understanding of all the vendors used in the REMS program as well as their roles and responsibilities.

References

FDA (2022a) Food and Drug Administration Compliance Program (7346.832) Chapter 46-New Drug Evaluation: Preapproval Inspections

FDA (2022b) Food and Drug Administration Compliance Program (7353.001) Chapter 53-Postmarketing Surveillance and Epidemiology: Human Drug and Therapeutic Biological Products: Post Marketing Adverse Drug Experience (PADE) Inspections Compliance Program for Human Drug and Therapeutic Biological Products

FDA (2023a) Code of Federal Regulations Title 21 Part 312.32: IND Safety Reporting

FDA (2023b) Code of Federal Regulations Title 21 Part 211: Current Good Manufacturing Practice for Finished Pharmaceuticals

FDA (2024a) Code of Federal Regulations Title 21 Part 820.22: Quality Audit

FDA (2024b) Rare Diseases at the FDA https:www.fda.gov/patients/rare-diseases-fd

International Conference on Harmonisaton (ICH) (2009) ICH Harmonised Tripartite Guideline Q10: Pharmaceutical Quality System

IQVIA (2018) Orphan Drugs in the United States: Growth Trends in Rare Disease Treatments

Chapter 5
Audit Preparation

In this chapter, we will explore the activities the PV auditor needs to consider while planning for a PV Audit. Each PV audit is different and will have unique requirements based on sponsor requirements. Of note, there are variations in how an auditor will prepare and conduct an audit. There is no one correct style or approach to effectively prepare for an audit. Each auditor will have their own unique method of audit preparation.

5.1 Audit Location

With the onset of the Coronavirus (COVID-19) pandemic in 2020, the Pharmaceutical industry needed to find ways to ensure regulatory compliance while adapting to the changing work environment. As a result of the pandemic, many employees needed to work from home and, as a result, PV audit activities needed to continue. The traditional audit model, which includes conducting an on-site audit at a client or vendor location, have face to face interviews, review the documents and data while on-site then write the audit report remotely was no longer a feasible option. Alternative methods of auditing needed to be developed. Thus, companies needed to become proficient in hosting remote PV audits and the PV auditors needed to become very skilled at conducting such audits.

PV professionals are very much aware that their typical daily tasks are focused on process development and assessment as well as data analysis. All of these are managed and retained electronically. It is rare that any PV information exists in a hard copy paper form, although this can sometimes occur. In my experience, the transition from on-site to remote PV audits has not been a significant change for many companies. The changing audit model did result in companies needing to refine existing information technology infrastructure in terms of applications and security associated with online meetings, document exchange platforms, etc. The

H. Murphy, *A Guide to Pharmacovigilance Audits in the USA*,
https://doi.org/10.1007/978-3-031-73596-7_5

bigger challenges resided with smaller companies that were not familiar with the operational requirements needed to host remote audits.

Change can be difficult, but there can also be advantages and efficiencies. As of the writing of this book, all of the PV audits I have conducted since March 2020 (~160 audits) have been remote. In fact, I have not received requests for on-site audits since the beginning of the pandemic. That is not to say that there will not be any, I have just not received any requests. My point is that the paradigm shift from the traditional audit model to a remote PV audit model brings with it many efficiencies. The audit sponsor does not need to incur travel expenses for the auditor, meetings can begin and end on time, and subject matter experts (SMEs) can attend their assigned audit sessions and return to their daily work when the session has ended but can be available if needed at a later time. No longer do audit attendees need to sit in a packed conference room for several days in case they are needed, and auditors can focus on the discussions at hand rather than be constantly interrupted by audit attendees coming and going from the conference room and having side conversations which ultimately distract the auditor. The auditor can now often (but not always) review documents in advance of the audit so the audit conduct can be focused on interviews and discussions of agenda topics rather than document review, although there will always be an element of document review during the audit. In addition, the auditor no longer has to monitor their time so they can conclude the audit in order to catch a flight. Some will counter my position by suggesting that face-to-face audits are preferred as you can more effectively develop a rapport with the auditee. I disagree as I have effectively done this many times over via online meetings. I suppose this comes down to individual preference but, from a strictly objective PV perspective, remote audits (to me) are very effective and efficient.

In fact, according to Davies (2024), a recent German study showed no difference in the efficacy and effectiveness of remote audits when compared to traditional on-site audits. Depending upon the auditor's approach, remote audits may require a little more preparation than on-site audits since the auditor is frequently required to develop a preaudit document request list; once these documents are provided, ensure review of these documents in advance of the audit dates. Given the provision of this information in advance, I have found that this specific exercise allows me to be more prepared prior to starting the audit. I have more information which always allows me to focus my audit strategy and tailor questions based on the receipt of this information.

Remote audits offer more flexibility than on-site audits (Colman (2023)). Traditional on-site audits require auditors to manage schedules around travel, and auditees need to ensure the appropriate SMEs are available on-site on specific dates. As many companies have more than one location, some of these SMEs will need to travel to the audit site. This is an additional cost for the auditee. Remote audits permit personnel to efficiently use their time and ensure the right individuals have access to the appropriate processes. This improves efficiency for global teams located at multiple sites as they do not have to physically travel from site to site to participate in an on-site audit.

5.1.1 Auditor Tip

Ultimately, the sponsor, auditee, and auditor will need to determine whether the audit will be on-site or remote. I do recommend, if the sponsor requests an on-site audit, that the auditor ask the reason an on-site PV audit would be needed. It could be that there is more information that has not yet been shared with the auditor requiring the on-site audit. It could also be that the sponsor may not be used to remote audits and is not aware of the advantages of PV audits. Regardless, having open communication with the sponsor is important. As an auditor flexibility is key, the sponsor and auditee can benefit from your experience and expertise.

5.2 Determining Audit Purpose and Scope

The purpose and scope of the Pharmacovigilance (PV) audit is typically determined by the sponsor; however, it is important to have an open discussion with the sponsor to determine what the objectives of the audit are. The sponsor may initially have certain objectives in mind, but once you and the sponsor discuss the upcoming audit in terms of purpose and scope, this may change. In fact, the purpose and scope varies depending on whether the audit is for a business partner, vendor, or internal PV audit. It can also vary given the contractual requirements of a vendor and the number and type of products in scope.

5.2.1 Audit Purpose

The purpose outlines the reason for the audit. The audit could be a routine vendor or business partner audit, for-cause audit, or internal audit. The purpose outlines why the audit is being performed (e.g., "This is a routine surveillance audit of vendor "A" contracted to provide postmarketing Pharmacovigilance services for product "B" in the USA"). There are two main reasons to conduct an audit: compliance and ongoing improvement. The sponsor wants to ensure the auditee is compliant with regulatory requirements, contractual obligations, procedures, and best practice standards. If the auditee is noncompliant, then remediation is required. The remediation can take the form of a Corrective and Preventative Action (CAPA) plan where the identified nonconformances are outlined and immediate corrections (Corrective Actions), investigation as to why the nonconformance occurred and measures to prevent them from happening in the future (Preventative Actions) are documented and implemented.

5.2.2 Audit Scope

The scope of the audit provides more detail in terms of topics to be covered during the audit. The scope should be based on the nature of the business activity being audited, the complexity of the business activity being evaluated, the risk associated with the business activity being audited, and the impact that potential noncompliance could have on the sponsor. The scope does not include every agenda item but does provide an overall summary of general topics to be discussed. For example, "The scope of the audit will include the quality management system, project management activities, Adverse Event (AE)/ Product Quality Complaint (PQC) management, signal detection, literature search activities, aggregate report development and submissions, and safety database hosting and management."

5.2.3 Auditor Tip

It is often valuable to have a preaudit discussion with the sponsor to determine the purpose and scope of the audit as well as address any questions you have. This discussion will provide an opportunity for you and the sponsor to discuss the purpose and scope and discuss any areas of concern that need additional focus. I recommend that you use this meeting to ask any/many questions as needed to ensure a clear understanding of what the sponsor is expecting from this audit. You can provide suggestions from prior experience that may allow the sponsor to achieve their audit goals. I have always found that many auditees, although they may not expressly tell you, are looking for your advice and guidance. This also helps to develop a positive and collaborative communication with the sponsor.

5.3 Assembling an Audit Team

The selection of the correct audit team is critical to the overall success of the audit. An audit team will include a lead auditor who will manage all aspects of the audit team. The lead auditor will then be responsible for selecting the additional audit team members as needed. The selection of the audit team should focus on audit process competency but also on subject matter expertise, communication skills, analytical skills, and the ability to work within a team. Excellent auditors may have subject matter expertise in audit methodology and other areas; however, they may not have other necessary skills (e.g., communication skills, ability to work within a team, etc.) which are equally as important for the overall success of the audit. The assembly of the audit team requires planning and strategic evaluation of the audit purpose and scope as well as the expertise and additional skills needed to execute an effective audit.

5.3.1 *Lead Auditor*

The lead auditor has the responsibility for the overall management of the audit. This individual needs to be an experienced and competent auditor but must also know how to select and manage the appropriate team members with the necessary skills to contribute to the audit. The lead auditor has various responsibilities including informing the auditee of the purpose and scope of the audit; preparing the audit plan, audit agenda, and preaudit document request list; managing all audit team meetings; conducting the audit opening meeting; managing and ensuring the audit is being conducted appropriately in line with the agenda; acting as the primary contact between the sponsor and the auditee; conducting daily briefings during the audit; reviewing the findings and observations from the audit team; conveying the observations to the auditee during the close out meeting; managing audit team dynamics; and developing and submitting the audit report to the sponsor.

The lead auditor is responsible for the selection of each team member and therefore responsible for ensuring each auditor understands the scope of their responsibilities throughout all stages of the audit process. This should be defined during the audit planning process in advance of the conduct of the audit. Such planning will present a professional and organized approach to both the auditee and sponsor.

It is common that some companies may only require a single auditor to manage all aspects of the audit. In this situation, the selected auditor must determine if they have all of the skills needed to achieve the purpose and scope of the audit. For example, some PV audits do require the auditor to have a solid knowledge of computer system validation, computer system security, operational controls, and audit trails as required by FDA 21 CFR Part11: Electronic Records, Electronic Signatures. If the auditor does not have such knowledge, you must inform the sponsor of this during your preaudit call. The ultimate goal is the success of audit. If the auditor does not have the necessary skills needed, the auditor is responsible to convey this to the sponsor and provide suggested alternative auditors for their consideration.

5.3.2 *Audit Team Members*

Audit team members are selected based on auditing experience and subject matter expertise, but there are other characteristics that are equally important. The composition of the audit team should be aligned with the expertise needed to achieve the audit objectives and the complexity of the audit. Some areas to consider include:

5.3.2.1 Education and Experience

The PV auditor should have a solid understanding of PV principles, regulations, operations, and analytics. This will vary based on the scope of an individual audit. Some companies require that PV auditors have a designated audit certification, some require PV background, and some require both. In my experience, I have learned that PV is a very nuanced and regulated discipline. Understanding effective audit methodology and principles is helpful, but the understanding of how PV departments and teams work daily is equally important. The combination of audit methodology and PV work experience, in my opinion, is the most effective combination. Each auditor should be selected based on the individual strengths that the individual will bring to the audit team. One may have extensive audit experience, while another may have a depth of PV understanding. If the necessary competence is not covered by the auditors in the audit team, technical experts with additional competence should be made available to support the team. The selection of these individuals will strengthen the overall audit team.

5.3.2.2 Personal Characteristics

An auditor's personal characteristics are just as important as technical competence. Some characteristics to consider include patience, tenacity, resourcefulness, diplomacy, maturity, ethicality, calmness, open-mindedness, organized, systematicness, independent, adaptability, unbiasedness, positivity, and curiosity.

5.3.2.3 Communication Skills

An auditor's communication skills allow the auditor to effectively obtain needed information to evaluate the validity of an auditee's actions. Some important communication skills include (but are not limited to) listening, paraphrasing, handling disagreements, speaking, and writing in a clear and concise manner, appropriate use of body language, the use of open-ended questions and the ability to be tactful. Versatility is key.

5.3.3 Auditor Tip

The lead auditor is responsible for identifying subject matter experts with the applicable skills to ensure the success of the PV audit. Do not assume that because someone is a technical expert they have additional skills (e.g., communication skills) to work with an auditee. You must evaluate each team member holistically. Ultimately, the success of the audit is dependent on the assignment of the right person for the job.

5.4 Management of Audit Communications

Audit communications are between the sponsor, auditor, auditee, and audit team. As the audit team leader, the lead auditor is responsible for ensuring these communications are initiated, ongoing, and successful. In the early planning stages, communications are usually limited between the sponsor and auditor until the audit purpose and scope is determined and the audit team is selected. A single point of contact between the sponsor and auditee is optimal, which is part of the lead auditor's role.

The preaudit meeting with the sponsor is an opportunity for the lead auditor to develop a rapport and for the sponsor to develop trust and confidence in the lead auditor. This provides the auditor an opportunity to display a systematic and professional approach to the audit. The discussion and inquiry into the sponsor's audit process, conveyance of the lead auditors' approach to the audit, and achieving buy-in of the audit approach from the sponsor is important at this stage. It is valuable to show the sponsor that, as the lead auditor, you are flexible, thoughtful, and thorough in your audit approach. This will provide them with a sense of confidence in your abilities and knowledge that the audit will be conducted as expected. The auditor should establish not only the primary point of contact but also the primary method of communicating (e.g., email, phone call, online meetings, etc.). This concept should also be extended to the auditee.

As part of the audit planning stage, determining how documents should be exchanged between the sponsor and auditor as well as the auditee and auditor needs to be established. The lead auditor should determine if the sponsor and auditee have the capability to provide the audit team access to an electronic document exchange platform (primarily for remote audits) as some small companies do not have that internal capability. If the auditee does not have this capability, the lead auditor can follow up with the sponsor to see if they can provide this capability for the auditee. If sponsors do not have access to an electronic document exchange platform, there are several online options they can consider, but this would need to be determined internally by the sponsor.

Communication during the conduct of the audit involves a variety of factors including active listening, working with auditee escorts, working with translators, managing audit team communications, conducting daily audit briefings with the auditee, utilizing methods of technology to improve communications, and accommodating changes to the audit schedule. It should be mentioned that although the lead auditor is ultimately responsible for audit communications, the entire audit team is responsible for ensuring audit communications are clear and consistent. Audit team members are responsible for identifying and informing the lead auditor of potential issues for evaluation.

Upon completion of the audit, the lead auditor is responsible for the provision of the audit report to the sponsor including revisions as needed. Some sponsors will request that the lead auditor manage and review the CAPA plan with the auditee to ensure the auditee is meeting the expectations of the auditors' observations.

5.4.1 Auditor Tip

The development of positive and constructive communication with the sponsor and auditee cannot be overly emphasized. Both parties need to see that you, the auditor, are working toward the ultimate goal of ensuring audit success in a constructive, collaborative, non-biased approach. The sponsor should be kept informed at all stages of the audit process if issues arise with the auditee that could jeopardize audit success. The sponsor should be notified as soon as possible so they can determine the best course of action moving forward.

5.5 Developing the Audit Plan

The need for an audit plan will be determined by the audit sponsor. Not all audits require an audit plan, and the need for this document is dependent upon the sponsor's internal audit program. Some companies do require an audit plan and an audit agenda, some require a combined audit plan and agenda, and some only require an audit agenda (see Appendices for examples). The audit plan provides the framework for how the audit will be conducted including the regulations and document standards by which audit activities will be measured. The audit plan content varies but can include sections for the client contact information, audit scope, audit purpose, products to be included within the audit scope, regulations to be assessed against, contractual obligations, internal procedures, and vendor procedures. The audit plan will need to be reviewed and signed by the sponsor. Depending upon the sponsor's requirements, the audit plan will be provided to the auditee by either the sponsor or the lead auditor.

Audits are conducted to ascertain the validity and reliability of the information provided and to provide an assessment of the control of a system. It provides the management with information on the efficiency with which the company controls the quality of its processes and products (Agarwal and Mishra (2019)).

The primary goal of an audit is to independently and objectively evaluate existing activities to determine if they meet the established standards. The effectiveness of a PV audit is measured against the predefined standards outlined in the FDA regulations (for the purposes of the scope of this book), internal procedures, contractual obligations, and additional documentation outlining the roles and responsibilities of those performing specific activities on behalf of the sponsor. The audit plan developed by the lead auditor is usually one of the first audit documents to be created. It is developed based on the purpose and scope of the audit.

5.5.1 Auditor Tip

The auditor should ensure all elements of the audit plan are agreed prior to the development of the agenda. The review of the audit plan provides the sponsor the opportunity to make audit scope changes if needed, which can directly impact the audit agenda. Always confirm which party (auditor or sponsor) will provide the audit plan to the auditee. I have encountered several companies who did not share the audit plan with the auditee as it was strictly an internal document.

5.6 Development of the Audit Agenda

The audit agenda is the document that outlines the topics that will be discussed during the audit. I have seen different approaches and methodologies used for the development of an audit agenda. All are effective for specific audit types. The auditor should choose the best option to provide organization, structure, and clarity for the audit.

The development of an effective audit agenda requires an understanding of the areas of focus, the associated topics, and discussions that will allow the auditor to obtain the most effective and accurate information. The provision of sponsor and auditee contracts is a good starting point. This provides information on the contracted services which is typically the basis for most vendor audits. A company overview including the organizational structure relevant to the audit and inspection history is included. The overview of the quality management system (QMS) is usually included as it provides foundational information of how procedures and processes are controlled and managed.

Specific to PV are Pharmacovigilance Agreements (PVAs)/Safety Data Exchange Agreements (SDEAs) and other documentation that delineates the performance of PV activities. Previous audit reports with CAPA information is very useful as they can provide a snapshot of strengths and weaknesses identified during prior audits. In addition, some companies request a review of all CAPAs that resulted from the prior audit added to the agenda (even if they are closed) as confirmation that all planned remediation activities were indeed completed. A section on the review of deviations/CAPAs since the last audit based on a selected sample (or for the past 2 years whichever is earlier) is usually included in a PV audit agenda. In addition, sections detailing the contracted services including AE/PQC management, medical information call center activities, literature search and analysis, signal detection, aggregate report management, project management activities, risk-benefit analysis, regulatory submissions, and computer system and applications (to name a few) used in in support of the activities in scope may be included in the audit agenda.

Below is a description of the most common PV audit agenda formats that I have seen used. As I mentioned previously, there is no one best format. The auditor needs to determine the best format for the specific audit. See Appendix 12.2 for examples.

5.6.1 *Detailed Format*

This audit agenda format is extremely detailed and provides bullet points on each topic and subtopic to be discussed, the SMEs to discuss each topic, and the time allotted for each discussion. I have seen these agendas run 10 to 15 pages long. The amount of detail can be advantageous for some auditees but anxiety producing for others. Some auditees may have concerns about SME availability and time commitment. This agenda format tends to provide less flexibility for both the auditee and auditor.

5.6.2 *Limited Format*

The limited format agenda is very high level with limited information. Such agendas are less than one (1) page in length even for audits that are multiple days in length. This format lists high-level topics (no subtopics) for discussion and does not include areas for SME assignment or discussion timeframes for each topic. Such agendas provide little structure for both the auditee and the auditor and are rather free-flowing.

5.6.3 *Moderate Format*

A more moderate approach (which I tend to use) is an agenda that assigns proposed topics for discussion, session timeframes, and SME information that will speak to individual topics. This format contains enough detail to be informative and sets expectations for each day of the audit. These agendas can be 2 to 3 pages (per audit day) and longer depending on the length of the audit.

My preference for this agenda is rooted in prior experience. I needed to determine how I could inform the auditee of the audit topics to be discussed and provide an estimated timeframe for each session while maintaining organization and structure but at the same time continuing to move the audit along so we could conclude on time. That was when I decided to modify the amount of audit content details in the agenda. I experimented with the two other examples described above but found that the moderate format option worked best for me while auditing.

The detailed format example provided so much detail that it created anxiety and a certain amount of panic for some vendors. They were overwhelmed with the amount of detail to be covered in the audit and did not know how to adequately prepare for the audit.

The limited format example was so lacking in specifics that did not provide enough structure or detail for me, the audit team, or the auditee. As a result, the audit became (in my opinion) rather free-flowing where some SMEs attended when needed, others did not show up and the audit team could not conduct effective interview sessions as topic discussions were fragmented, and SMEs did not know when to attend.

In my experience, developing an agenda that provided a list of topics to be discussed and then integrating additional subtopics during the audit discussions provided me with the opportunity to obtain the information I needed for a topic. Providing the estimated session timeframes allowed the auditees to manage SME attendance and balance other work commitments. It is important that the audit agenda is reviewed by the sponsor prior to sending it to the auditee for review.

5.6.3.1 Auditor Tip

Each auditor must determine what audit agenda format works best for them to assist in the completion of audit objectives. The sponsor may request you utilize their audit agenda template, but you can modify the template to include the pertinent information you think is needed. Most sponsors (in my experience) are very flexible in the modification of their templates as long as the overall objectives of the audit are being met. Ultimately, the agenda provides the structure for the audit and is not meant to be all inclusive. You may see some auditees provide modifications to the agenda where some topics are removed. Always follow up to inquire as to the reason such changes were made.

5.7 Preaudit Document Requests

As I previously mentioned, the many PV audits are now remote. As a result, auditees need to utilize secure document management platforms for the secure and controlled exchange of preaudit documentation. There are many options available, and all are great options. The choice of the platform depends upon the organization's preference. In preparation for the audit, the sponsor will typically provide applicable Master Services Agreements (MSAs), Statement of Works (SOWs), Safety Data Exchange Agreements (SDEA), Pharmacovigilance Agreements (PVAs), and other contractual agreements. Such documents provide the initial framework for audit planning.

It has been my experience that a majority of auditees are now very well-versed in conducting remote audits and will provide preaudit document requests from the auditor via a secure document platform. There are still some vendors who are resistant to providing any documents in advance of the audit citing internal policies. In my opinion, the auditor receipt of preaudit documents improves the overall conduct of the audit allowing for the auditor to have a baseline by which to prepare for and execute the audit. Entering into an audit without any reference point by which the vendor is organized, what documents exist to support contracted activities or computer systems can negatively impact the overall conduct of the audit. In essence, you are playing "catch up" before the audit even starts. In fact, if this were an inspection, the FDA inspector would literally have more information on this auditee than you would as an auditor. In addition, if documents are not provided in advance of the audit, the auditee will still need to provide them "via screenshare" during the audit. In the instance of SOPs or WIs, you will need to endure the auditee controlling the document and advancing the document as you read it on screen. This is not efficient and takes time away from interview sessions as well as impacts your ability as the auditor to adequately evaluate the process against contractual and regulatory requirements. I have found the approach of requesting targeted preaudit document requests to be immensely successful. In fact, for those auditees refusing to provide any pre audit document requests, I have informed the sponsor just to keep them informed. Sponsor understands the importance of the auditor having this information in advance of the audit and will often advocate on your behalf. See Appendix 12.3 for examples.

The type and number of preaudit requests will vary based on the type of PV audit to be conducted. Some of the documents that can be requested include, but are not limited to:

- Preaudit questionnaires.
- Organizational chart.
- Full list of auditee SOPs, WIs, and policies.
- Auditee procedures for:
  - Quality manual (if applicable).
  - Controlled document management.
  - Training program.
  - Deviation and CAPA management.
  - Internal audit program.
  - Vendor selection and management.
  - Record retention.
  - AE/PQC corporate policy.
  - SOP for AE/PC handling and management.
  - IT system security.
  - IT change control.
  - IT system monitoring and maintenance.
  - Business continuity.
  - Disaster recovery and evidence of most recent testing.
- List of all personnel for the sponsor project (usually for vendor audits).
- List of Adverse Events (AEs) since the previous audit (for vendors involved in Individual Case Safety Report (ICSR) management).
- List of Product Quality Complaints (PQCs) since the previous audit (for vendors involved in PQC management).
- List of combined AE/PQCs since the previous audit (for vendors involved in PQC management).
- List of deviations and CAPAs since the previous audit (usually for vendor audits).

- Auditee compliance metrics.
- List of computer systems and applications used to support PV activities.
- Reconciliation reports (specify timeframe and vendor (there may be multiple vendors)).

5.7.1 Auditor Tip

Whether an auditor wants to request preaudit documentation is entirely up to the individual auditor. This is by no means a requirement. I find this practice useful for me but may not be for others. Some auditees do not have access to a document exchange platform. If they do not, ask the audit sponsor if they can provide restricted access to their platform for the duration of the audit. Most sponsors are happy to do this.

5.8 Technology

There are currently numerous methods by which auditees can securely exchange preaudit document requests. Some companies have their own document exchange platforms internally, while others utilize third-party platforms. In some cases, the client can provide the auditee limited access to their platform for the purposes of the audit. Ultimately, the auditor should ensure that the client and auditee are comfortable with the method by which their information is being exchanged.

Organizational procedures should specify preferred options. The auditee and sponsor should choose solutions that allow the audit objectives to be met. These tools do not have to be cost prohibitive, be complicated to use, or have numerous functions. They should be fit for purpose. Data security and accessibility are a balance. Each company has valid cybersecurity concerns which they need to consider on behalf of their organization. Security measures should not restrict business practices. The company should consider a backup solution should the primary file-sharing solution be unavailable. With the use of various technologies, the auditee should consider the following: should images be recorded or provided in real time, is there a need for interviews to be recorded and/or transcribed, the security features and vulnerabilities of the selected platform, and how can the information be accessed and secured for multiple parties. As part of this decision, auditees need to consider how to navigate technological issues with auditors, business/license partner, and vendors.

The common use of standard cloud-based document sharing solutions can meet most companies' standard security requirements. Some have the capability for transcription and recording but that does not mean these functions should be used as other participants may have their own security requirements. These requirements can include organizational cybersecurity practices, industry cybersecurity best practices, and regional compliance regulations such as the EU General Data Protection Regulation (GDPR), to name a few. During the selection of such a platform, the auditee should consider security measures such as follows: is encryption available for video conferencing, has a GDPR assessment of the platform been performed (you can have virtual participants from the EU in an FDA inspection), if video recordings are planned, have all participants provided consent for the recording (if not how is participation for these individuals managed), secure storage, access to and retention of such recordings, the location of the servers, and the ability of third parties to access this information.

5.8.1 Auditor Tip

The auditor should discuss the method for preaudit document exchange with the sponsor during the preaudit meeting. Some sponsors will use a platform to provide the auditor with contracts and additional documentation, while the auditee will use their own platform to provide additional preaudit document requests. Some sponsors will require that they establish a single platform by which all audit documents are exchanged and stored by the sponsor and auditee. Establishment of platform access can take some time depending on the organization's business practice and processes. It is best to begin this discussion early in the audit process.

References

Agarwal P, Mishra A (2019) Pharmaceutical Quality Audits: A Review, International Journal of Pharmaceutics, ISSN, Vol 11, Issue 1

Colman K (2023) The Future of Virtual Audits and Audit Success https://www.bioprocessintl.com/analytical/qa-qc/the-future-of-virtual-audits-and-audit-success/

Davies J (2024) Pharma's remote audits are here to stay: Distinctive strategies for doing remote audits successfully https://www.pharmamanufacturing.com

Chapter 6
Audit Conduct

6.1 Audit Conduct

In this chapter, I will describe the primary activities included in the conduct of PV audits. I will discuss a typical audit approach and content to be covered and provide insight as to what elements should be included in each section. A reminder that this is merely an overview based on my professional experience. As I have mentioned before, each PV audit is unique and should be approached as such.

6.2 Audit Logistics

Audit logistics frame when and where the audit will be performed, the participants involved, and what technology is needed. Logistical considerations include audit location, audit attendees, time zones for participants, geographical locations, cultural considerations, and technology to be used for remote audits and technological capabilities of the auditee and various participants. Audit logistics should be addressed well in advance of the audit conduct but is a consideration throughout all phases of the audit process. Addressing this audit element can be achieved through ongoing communications with the sponsor and auditee. As previously mentioned, some smaller companies may not have the technology needed to successfully host a remote audit so they may require additional assistance.

Let's discuss these logistical considerations in more detail.

H. Murphy, *A Guide to Pharmacovigilance Audits in the USA*,
https://doi.org/10.1007/978-3-031-73596-7_6

6.2.1 Audit Location

The location of the audit (remote versus onsite) will require specific considerations. The audit location is determined during the audit planning phase. Regardless of auditor preference, each auditor needs to consider the scope and objectives of the audit and determine whether a remote or onsite audit would provide the best chance of success. On-site audit considerations may include the provision of meeting rooms for audit team use, dedicated office space for auditors to conduct document review, safety, environmental or security issues, high speed internet, and other means to communicate with the auditee. Whether or not an onsite audit is feasible for team members located in different part of the United States or globally should be examined. For example, if you have an auditee office location located on the West Coast of the United States but the majority of the audit SMEs are all located on the East Coast, would it make sense to have all SMEs travel to West Coast office location for the audit or might it be prudent to conduct the audit remotely so the SMEs can attend for their sessions then continue their daily work activities. Another scenario might be the audit of a Specialty Pharmacy, where the inventory and dispensing activities are part of the PV audit scope. In this situation, an onsite audit may be suggested as it is difficult (but not impossible) to view the inventory and dispensing process via a remote meeting platform. The geographical locations of audit participants, including the impact of requiring SMEs to travel for the audit, should be considered by the sponsor and the auditee.

There are issues for both remote and onsite audits to consider including access to the necessary facilities and equipment required for verification activities, access to documents and records, management of proprietary and confidential information, and permission to use cell phones cameras or recording devices. Many of these activities are the responsibility of the auditee, but the capability to manage these issues will impact whether the audit should be remote or onsite.

6.2.2 Audit Attendees

Audit attendees should be strategically aligned to provide the information required by the auditor and allow for efficient and effective conduct of the audit. Additional attendees that are not SMEs or involved in the hosting or conduct of the audit should be limited. Some sponsors and auditees will have additional employees attend for learning purposes, but this can be disruptive, particularly for onsite audits. The auditor should be aware of who will be attending each session.

There are confidentiality issues to consider given the sensitive nature of audit information discussed and presented as evidence. In general, audit participants should consist of the auditor, auditee host, auditee scribe, auditee document request manager and the SME(s) needed for a specific audit session. Some auditees take the position to have 15–20 people attend (onsite and remote) for every session just in case someone is needed for a random question. This is not an efficient or effective approach to running an audit. Unfortunately, with this many people involved, there are often numerous interruptions and distractions including, ongoing conversations not relevant to the audit, and participants are not engaged as they are completing their daily work activities and when called upon for a question do not know the subject of conversation or what question has been asked of them. This requires the auditor to backtrack, repeat the questions, and interrupt the ongoing interviews.

The lead auditor is responsible for ensuring the effective and efficient conduct of the audit and as such has the right to limit the number of individuals participating if there is the potential to negatively impact the audit conduct. I have on several occasions spoken to the audit host to request that only the SMEs needed for a specific session should attend due to the countless distractions I need to address, which impact and ultimately delay the audit conduct. This may not be received by the auditee in a positive light; however, the success of the audit is of paramount importance. The auditor needs to control the situation diplomatically but firmly. If a question arises during a specific session that the current SME cannot answer, the question can be directly asked by the auditee audit team to the correct individual. An answer or additional interview can be conducted subsequent to the current session. This eliminates the need for having 15–20 people to attend every audit session. Audits move quickly and require the auditor to not only have excellent audit and communication skills but the auditor also needs to utilize critical thinking and problem-solving skills throughout the entire audit. The auditee's responsibility is to provide the best environment for the auditor to do their job. The most successful and efficient audits I have conducted consisted of no more than 3–4 people attending at any one time.

This concept also applies to attendees from the sponsor. The auditor should understand why the sponsor needs to attend this independent audit and who the attendees will be. It could be that based on the sponsor's audit program, representatives from the sponsor audit team are required to attend. It can also be that the sponsor wants employees to attend for learning purposes. The auditor is not responsible for training anyone participating in the audit on how to conduct an audit. If the sponsor or auditee want employees to attend for training purposes, they should confirm with the auditor if it is acceptable for them to attend.

6.2.3 Time Zones

It is no surprise to any auditors reading this book that managing numerous time zones where SMEs are located can be a challenge. Given the increased frequency of remote audits versus on-site audits and the broad geographical locations of all attendees, this can be a greater challenge. The auditor, auditee, and SMEs (and sponsor if attending) may be located in different time zones. The auditor needs to remember that you are relying on the auditee to be available to provide you with the information needed to conduct the audit on behalf of the sponsor so flexibility is key.

The pharmaceutical/biotechnology is truly global in nature. A company may have the office headquarters located in the United States, but employees and vendors can be located in various parts of the world. This is a consideration for even a US-based FDA audit. Just because the company has its headquarters in the United States does not necessarily mean that timings for the audit will be based on US time. Even within the United States, there are nine standard time zones covering the states, territories, and other US possessions. According the World Atlas (2024), the US time zones include Pacific, Central, Mountain, Eastern, Alaska, and Hawaii. The other three time zones include the Atlantic Time Zone, followed in Puerto Rico and the US Virgin Islands; the Chamorro Time Zone, followed in Guam; and the Samoa Time Zone, followed in American Samoa. This does not include any other global time zones where additional auditee or sponsor attendees may be located. Clearly, not all time zones are involved in each audit, but this illustrates the complexity, organization, and flexibility of attendees that need to be considered.

There are several ways to approach the complexity of various time zones. All are acceptable options provided the key individuals are included at the appropriate time.

- Sponsor Time Zone

If the audit sponsor is attending the audit one approach would be to have the primary time associated with the sponsor location. Since the sponsor is the party requesting the audit, they are responsible for ensuring that the audit is conducted in line with the audit plan, contractual obligations, and regulations. They also have global knowledge about how the auditee's activities can impact other parts of the global PV system. This information is typically shared with the auditor ahead of time but not in all instances. The sponsor may have additional concerns or questions to address with the auditee that may not be known by the auditor. I rarely see this approach used from a timeframe perspective, but it is not unheard of.

- Auditor Time Zone

Some auditors will insist that their time zone be the primary area for the scheduling of audit sessions as well as management of the audit. Since the lead auditor is the audit manager in addition to the auditor, the thought for some is that this should be the primary time zone used and all other participants need to accommodate this time. For a US-based audit, this becomes difficult to manage if the auditee or auditee's vendor is located in the Asia Pacific (APAC) area. In my opinion, this approach is not particularly flexible as this would require (based on locations) SMEs to attend sessions well into the late evening and early morning in their local time zone. For example, if the auditor is starting an audit in the United States at 9:00am EST and the auditee is located in the Asia Pacific area, the audit start time for this auditee could be 9:00pm (or later) local time. This would require these SMEs to attend sessions through the late evening and early morning hours. This does not show flexibility or consideration for the SMEs location.

- Auditee Time Zone

Scheduling audit sessions based on the auditee or auditee vendor time zones ensures that the required SMEs will be available to attend the scheduled sessions. This will require thc auditor to be flexible as it is very common to be conducting a PV in the early hours of the morning to accommodate SMEs in other parts of the world. This shows respect and a willingness to work with the auditee. This can go a long way in developing a positive rapport with the auditee. This is not to say that the conduct of the audit is directed by the auditee. This is not the case. There is a balance the auditor needs to maintain between being flexible and accommodating but also maintaining control of the audit. Theses session times can be moved around on the agenda to address SME locations.

This is the option I tend to use for many of my audits. Prior to the development of the audit agenda, the sponsor introduced me to the auditee as the auditor. While I am drafting the agenda, I will reach out to the auditee contact to confirm where the majority of the audit team will be located for this audit so that I can work this into the draft agenda. This provides the auditee the opportunity to consider the audit team, their location, and ability to attend specific sessions. Once the draft agenda is sent to the auditee, there is always opportunity to modify the agenda based on SME availability but at least this can serve as a starting point.

6.2.4 Cultural Considerations

Whether you are conducting an on-site audit or a remote audit, the lead auditor and audit team must always be mindful of cultural considerations of those attending the audit. There are numerous cultural norms, traditions and practices per region and country the scope of which cannot be covered in this book. The pharmaceutical/biotechnology is a diverse and global industry. US companies have office locations, vendor, and business/license partners across many geographical regions. As a result, the auditor may encounter SME availability issues due to national or religious holidays, etc. In some countries, the typical work week is not necessarily Monday through Friday as in the United States. Once again, flexibility is important. If audit SMEs cannot attend the audit on the selected days, this might require the auditor to communicate with the sponsor to discuss alternate audit dates, or it may simply require some shifting of audit sessions for certain individuals so they can attend on another audit day.

Onsite audits can be a little more complex as the lead auditor and audit team must be mindful of local cultural norms in terms of specific greetings, the auditee wanting to provide gifts, etc. Ultimately, you do not want to offend anyone, so the entire audit team should conduct research on local customs prior to arriving at the audit location. This shows respect and interest in the auditee and will go a long way in developing a positive relationship with the auditee and the overall success of the audit.

6.2.5 Technological Considerations

We have touched upon this a few times, but I think it is worth mentioning once again. In the United States, we are used to having access to the latest and greatest technological advancements. High speed internet, online meeting platforms, document exchange platforms, electronic document systems, etc. are the norm in the United States, but this is not the case in some areas of the world. There are some smaller organizations and geographical locations where access to these tools are limited or not available. The auditor should be aware of the sponsors' and auditees' technological capabilities. The auditor should not assume the auditee has the capabilities to provide documents in a secure manner during the audit. During the audit planning phase this is the time to discuss these capabilities with the auditee. If they are lacking in the technology needed for the audit provide options and assistance. The auditor can discuss the auditee limitations with the sponsor and the sponsor may be able to provide assistance. All parties (sponsor, auditor, and auditee) are working together toward the same goal so do not assume these discussions may have been had with the sponsor prior to the scheduling of the audit.

6.3 Audit Tools

There are various tools that an auditor can use to provide structure during a PV audit. Some are readily available on the internet, or the auditor can develop individual tools to aid in the audit organization. There are no right or wrong tools. In fact, some auditors do not use any tools. The auditor should use (or develop) the tools that work best to ensure appropriate audit organization.

6.3.1 Audit Checklists

Audit checklists can be used before and during the audit. The checklist contains a list of required items such as tasks to be completed or questions to be asked to the auditee along with required evidence to be reviewed. In many cases, the checklist is developed in line with the audit agenda sections, questions to be asked, areas to explore and evidence to request. The checklist also provides an area for the auditor to make comments on discussions or evidence presented. Audit checklists are flexible in that they can be tailored to the auditor's needs.

6.3.2 Audit Guidelines

Audit guidelines are similar to audit checklists but are usually more focused. For example, audit guidelines can be useful when reviewing auditee policies and procedures. These guidelines can be developed to document whether the auditee is conforming on nonconforming to required procedures, documentation, contracts, and regulations. These can be particularly useful for audit teams where the audit may be conducted over more than one location or across multiple departments within an organization.

6.4 Conducting the Audit

6.4.1 Opening Meeting

The opening meeting of an audit sets the tone for the overall audit. In my opinion, I think the opening meeting is critically important. The opening meeting should be focused and provide insight to the auditee as to how the lead auditor and audit team will approach the audit over the specified number of days. I try to keep the opening meeting brief (15–30 minutes). The meeting can be conducted utilizing methods such as PowerPoint slides or verbal discussion. Some sponsors require the audit team to use their slide deck templates for opening and closing meetings. Either method is acceptable. The key is to set the right tone for the audit.

Some common topics included in the opening meeting are (but not limited to):

- Introduction of the audit team and auditees.
  - It is helpful for all parties to understand who is attending and conducting the audit. For a PV audit, auditees want to see that the auditor and audit team not only have extensive audit experience but also understand the complexities and nuances associated with PV activities.
  - The auditor team should introduce themselves with a high-level overview of work experience emphasizing PV experience. You do not need to provide information on every company you have worked for in your career. The goal is to provide the auditee confidence that you have the expertise needed to conduct the audit.
  - Introduction of the auditees allows the audit team to understand if the appropriate individuals are participating based on the scope of the audit. It should be noted that not all SMEs typically attend the opening meeting. In my experience, I have found that the opening meeting usually has limited attendance or in some instances, may have many attendees that are not audit SMEs and may not participate in any other sessions.

- Review of the audit purpose and scope.
  - It is important to review the audit purpose and scope in terms of whether the audit is a qualification, routine, or for cause audit and the standards by which the audit will be conducted (e.g., US PV postmarketing regulations, vendor contractual obligations, PVAs/SDEAs). Do not assume that all attendees have knowledge of purpose and scope of the audit. I have noted over the years that this is not always the situation.
  - The audit team can provide a high-level summary of topics to be addressed.
- High-level review of the audit agenda.
  - Providing a high-level review of the audit agenda during the opening meeting is immensely helpful as it aligns all parties on the topics to be discussed during the course of the audit and permits the auditee to ask additional questions.
  - It is not necessary (or helpful) to review every bullet under each topics. Just hit the highlights.
  - Identify sessions that need to be adjusted and collaborate on the best time for that session.
  - Describe how the sessions will be conducted. For example, some auditors approach PV audits from a conversational standpoint. They review the procedures but then have the auditee explain the daily process, which often generates more questions by the auditor. Conversely, other auditors will review the procedures and ask questions for clarity based on what might not be clear in the procedure. Auditors should use whatever approach works for them.
- Confirmation of audit logistics.
  - This can include the audit team confirming who the primary contact will be during the audit for requests and questions
  - Confirm the primary method of communication between the auditee and audit team. It can be via email, via a messaging app, within the meeting platform, etc.
- Remind the auditees that the audit is being performed to assess compliance (discuss per specific audit), add value, and promote improvements not to specifically search for nonconformances.
- Remind the auditees that the audit is confidential.
- Remind the auditee that the audit is a two-way process, and they are free to ask questions at any time.

Auditor Tip

Auditors should keep in mind that, for some, this may be their first audit experience. As a result, these individuals may be nervous and apprehensive. Even for those that have participated in many audits there is still that sense of the unknown, which can put audit participants on guard. This is especially true if they had a prior negative experience. A skilled auditor will quickly recognize these signs and it is their job to put these individuals at ease.

6.4.2 Audit Strategies

There are various audit strategies that can be utilized either during the planning phase or during the conduct phase of the audit. Some auditors may know these strategies by different names, and some may have used most of these methods without recognizing they were using them. I have briefly summarized the commonly used audit strategies I have used in my audit experience but there are many more not outlined below.

I have decided to discuss these techniques during the audit conduct section of this book as I often do not determine what technique I will use until the onset of the audit. This has been my experience, but auditors can use these or other techniques during the phase of the audit that is most appropriate for them. Such strategies allow the auditor to gain insight into processes and procedures and identify areas for improvement.

- Tracing Method

Tracing is a technique that lends itself to being used effectively during PV Operational audits. For audits that require an in-depth understanding of processes used, this strategy can be very effective. In my experience as a PV auditor, I find that reviewing existing PV procedures and then walking through the process with the auditee is very helpful in determining disparate areas between the procedure and the actual process performed. The tracing element comes into play when I have the auditee discuss the process from beginning to end, and interspersed within that, I ask a lot of questions. My intent is not to find fault with the process but rather to truly understand the process and prevent making assumptions. I am well aware, having spent a majority of my career working with PV, that processes are good at providing a baseline by which an activity is performed but do not necessarily provide for every potential scenario nor provide granular detail. Discussion of the process is very helpful in that regard.

Tracing itself is rooted in the principle of following a process in a chronological manner and collecting (and reviewing) objective evidence to prove whether the process is being performed as intended. Depending on where you begin, you can follow the process in chronological order from your starting point. Simply put, you can trace forward or backward. In forward tracing, the auditor begins either at the beginning or middle of the process and traces the process forward from the start point. Backward tracing involves starting either at the middle or the end of the process and tracing backward from the starting point.

- The Process Method

The process method can be used in conjunction with the tracing method or separately. The process method allows the auditor to examine sequential individual steps within a process along with the interrelationships between steps. An example of the process method can be the use of a flow chart to describe a process. This can be particularly useful in a process audit where the scope may include a single process or for use in a system audit. Flow charts are used commonly in PV and provide summarization and visual aid to complex processes. They depict sequential activities leading to an output. Auditors can use flow charts to develop an understanding of an overall process or system, identify points in the process that result in a decision, identify a role performing specific steps, identify potential problem areas that may impact other areas of the process, discover opportunities for improvements, and guide activities for CAPA management.

As per Russell (2013), the process method is useful to identify weaknesses within an organization's management system, securing appropriate inputs, breakdowns, and the achievement of output objectives.

- The Department Method

The department method allows the auditor to evaluate accountability within a department by focusing on the entire department rather than individual functions within a department by reviewing various elements within the department. An example of this method's use could be the evaluation of a vendors PV department.

- The Element Method

The element method refers to the assessment of individual elements within a system. This method can show a direct correlation between requirements and verification compliance. This could be used in PV audits where the sponsor needs to assess and evaluate a vendor's compliance with contracted PV services such as AE/PQC case processing, literature search activities, signal detection, and aggregate report management, to name a few.

- The Discovery Method

The discovery method allows to auditor to randomly audit areas that may lead to the identification and cause of a known issue. This method is also known as random auditing. This method would not normally be used for a planned audit but can be used when an auditor knows there is an issue, but the evidence provided does not clearly discern details of the issue. This method should be used sparingly but can be effective in unusual situations. You do not want the auditee to perceive you are searching for nonconformances.

6.4.3 Common Areas of Focus

Upon completion of the opening meeting, it is time to move on to the topics listed in the agenda. Each audit is unique but there are topics that are common to most US-based PV audits depending on contracted activities. They can include, but are not limited to:

- Organizational quality management system
- Project management activities (vendor audits)

- Adverse event/product quality complaint management (intake, processing, medical review, submissions)
- Translations
- AE/PQC reconciliation
- AE/PQC case review
- Overview of computer systems/applications used to support PV activities
- Literature searches and analysis
- Aggregate report development and submissions
- Signal detection
- Benefit risk evaluation
- PV quality management system

Since there is great variation in PV audits based on the audit types as described in Chap. 4, we will not discuss each potential PV topic but, rather, will focus on a few that are very common. I will provide insight as to how I have approached these sections during my audits. Again, there is no right or wrong way to approach these sections. This is merely my approach developed over time, which has allowed me to maximize the amount of information obtained from the auditees.

- Auditee Quality Management System

The auditee quality management system (QMS) sets the framework for the controls associated with document generation and retention, documentation standards and quality control within the organization. In my opinion, the QMS is the backbone of every pharmaceutical/biotechnology organization as it sets the standards for how activities are performed and controlled within the company. For just about every PV audit I perform, this topic is one of the first discussed. The QMS is an expansive topic, but I will focus on those elements that typically have the potential to impact PV on some level. My experience has shown me that many (but not all) identified PV issues can be linked back to weaknesses within the QMS. As a result, the QMS elements I will review with the auditee include document management, training, vendor qualification and oversight, CAPA management, and the company audit program. Let's review some of the most pertinent components of these topics to be explored.

Document Management

When I begin the discussion on this topic, the intent is to determine if a process exists to control and manage documents at the auditee. This includes whether the document management system is paper based (yes there are still some companies that do this) or whether it is managed within an electronic document management system (eDMS). Once this is determined, I will ask the appropriate SME to explain the process for the development of new documents versus updates to documents. This discussion may or may not be very detailed depending on the level of detail provided in the associated SOP (if there is one). In most cases, I really never see the level of detail discussed versus documented in these procedures. My intent is to understand how the documents are being initiated, reviewed, and approved. Even for auditees utilizing an eDMS, I have seen variations in their processes as one or several steps may be conducted outside of the system. It is important to understand how the auditee is managing the steps outlined in their procedure, where hand offs occur, and the controls implemented to ensure the organization is following the procedure. If they do not follow the procedure, how are these deviations being documented and managed? For those auditees using paper-based systems, how will the generation, review, and approval of these documents, including templates, sign off sheets, etc., as well as their storage and retention be discussed?

Other document management elements such as version control, access control, tracking of reviewers' comments and approval signatures will be explored. For those auditees with a robust eDMS, a user manual or guide is often created (or needed), so this will also be reviewed to ensure there is consistency in the use of the system. In addition, a review of compliance with 21 CFR Part 11 requirements will be requested and reviewed.

It is important that the auditor look at the overall organization being audited and set expectations appropriately. You cannot expect a company of 10–20 employees to have the same QMS infrastructure in place as a company of 1000 employees but there still needs to be measurable controls in place. A smaller company will most likely have a manual process, but it can still be effective and controlled compared to a large company with a robust eDMS where many activities are automated.

Training

The auditee training program is an important part of the QMS as it sets the foundation for how employees are trained when onboarded on an ongoing basis and on a project level. As an auditor, I will ask the auditee to discuss how training is determined prior to a new employee starting employment and on an ongoing basis. For example, I would discuss whether the auditee has a role specific curriculum, who within the organization determines which training an employee will receive and how the training is assigned and documented. In addition, if annual training occurs, who receives this training, how the training is provided, are there standard procedures or topics all employees or selected employees will receive and how this training is documented. Sponsor (or project specific) training will also be discussed.

It should be noted that, like document management, there are various ways of managing training assignments and documentation. Many companies utilize an electronic learning management system (eLMS) which provides for the assignment of training materials based on a predetermined role specific training curriculum. Such systems provide email notifications to employees and their managers of newly assigned training, reminders to complete training and notifications that training is past due. These systems also provide periodic notifications to managers of employee training status and can be used by the Quality department to monitor organizational training compliance and metrics.

Smaller companies may not have eLMS and manage the training via a manual process. It would be important to discuss the training curriculum (some companies may call it a training matrix), how training is assigned, performed, and documented. In addition, since there are no automatic reminders, understanding who and how the employee training oversight is performed would be considered. For example, if the employee-manager is responsible for ensuring the employee completes the training, the documented process for ensuring this is performed and documented should be discussed. If the training is past due, the auditor would want to understand the process for addressing late training.

For project-level training, it would be useful to understand if the sponsor conducted AE/PQC and product-specific training for the auditee, the frequency, and method of this training, how the training is documented, and if both parties agreed to a train-the-trainer model. A train-the-trainer model utilizes an approach by which specific auditee project personnel are trained by the sponsor on specific requirements. A common example of this in PV is training on AE/PQC reporting and product-level specifics performed by the sponsor with auditee personnel attending. Upon completion of this training, these auditee employees are responsible for training new project team members and the retraining (if needed) of existing project members. The retention of the training documentation from the sponsor and evidence of auditee training will also be reviewed. Some vendors, even though they normally utilize an eLMS for employee training, do not use the system for project-level training or training documentation. This would need to be clearly delineated as it is important to understand how the training, if not in the eLMS, is assigned (e.g., email), performed (e.g., read and understand or quiz), documented (e.g., form), and retained (e.g., auditee project folder on a secure drive or other location). The same questions would be raised for those companies with manual systems.

A sample of project team members training records, job descriptions and resumes/curriculum vitaes are typically reviewed. The reason for performing this as part of the audit is to verify that the individuals working on sponsors project have the appropriate work experience and education as outlined in the job description. This is part of the sponsors oversight activities. It should be noted that this activity is not always clear cut as some companies have their own policies that may preclude the auditor's ability to complete this task. Some vendors will not provide resumes or CVs to auditors citing confidentiality issues. I have run into this but only in a rare occurrence. This makes verification of work experience and education impossible. As a result, the sponsor can by informed and this can be noted in the audit report.

Also, some companies do not have job descriptions or require the employees to have resumes/CVs. This is a risk to the sponsor as there is no way to verify that the individuals working on their project have the qualifications to do so. Again, the sponsor should be made aware, and this can be noted in the audit report. In my experience, the vast majority of auditees readily provide all of this information as they understand the sponsor's requirement to verify that the project team members have the appropriate qualifications to work on their project.

Deviation and CAPA Management

The auditees deviations and CAPA management process has a direct impact on contracted PV activities. This process, as part of the auditees QMS, provides guidance by which the auditee will manage deviation and CAPAs that may occur during a sponsor's project. It is true that the sponsor can make specific requests for their project but whether the auditee agrees is based on the existing auditee policies and procedures. From a PV perspective, this is a vitally important part of the QMS. Deviation and CAPA management can be rather complex depending on how the auditee defines what constitutes a deviation, when a deviation requires and CAPA, how the CAPA process is performed and documented, and whether deviations and CAPAs that occur during a sponsor's project are shared with the sponsor.

A common situation in PV Operations is when an auditee, contracted to provide AE/PQC processing activities, was required to provide an AE to the sponsor within 1 business day of awareness, but instead, the case was sent to the sponsor two weeks following the initial receipt date. Obviously, both parties never want this situation to happen, but it does occur. The auditor will need to understand how this case was identified as late, how this was managed internally with the auditee, and on what date it was sent to the sponsor. Some vendors will automatically consider a late AE/PQC case to the sponsor as an event that automatically requires a CAPA, others do not. Determination of whether the late case is considered a deviation or CAPA can be based on the auditee's procedures, definitions of deviations and CAPAs, and frequency of occurrence. These factors will guide whether corrective actions are performed, whether an investigation is needed or performed, if and how preventative actions are implemented, and whether effectiveness checks are performed. The auditor will need to evaluate all of these factors against regulatory requirements to determine if the auditees' actions are appropriate. Ultimately, the important elements are to ensure that the deviation and CAPA effectively manage identified nonconformances, immediate corrective actions are implemented, and preventative measures are put in place to prevent recurrence. This is more nuanced and complex than one might expect.

Supplier Qualification and Oversight
A review of the auditee supplier qualification and oversight program provides foundational information as to how vendors are qualified and requalified. The auditee will need to qualify any PV vendors as part of their process. The information obtained in the review of this process provides the auditor with an understanding as to how vendors are selected. Most commonly, many companies use a risk-based methodology where the higher the risk of the vendor (based on activities) to the organization, the more robust the qualification process and the more frequent the requalification. In addition, it is helpful to understand how the qualification process occurs (e.g. audit questionnaire, full audit, or both), the cadence of ongoing audits and how a vendor may be deselected or discontinued.

Internal Audit Program
The auditees' internal audit program reflects the Quality department's involvement within the organization from the perspective of evaluating internal and external activities. Most companies will create an annual audit plan that outlines the internal functions to be audited as well as external vendors to be audited for that year. This plan is typically risk based where the higher the risk of the internal function or external vendor to the organization the more frequently (often annually) the audits will occur. This program usually defines how departments and vendors are selected for audit, the frequency of these audits, whether internal or contract auditors will be

performing the audits, and how observations identified during the audits are managed. PV is usually part of the annual audit program where internal PV departments or specific functions within PV may be audited along with specific PV vendors. In addition, many mid- to large-sized pharmaceutical/biotechnology companies will have their own audit plan, which should be aligned and shared with the Quality function.

- Project Management Activities

The review of project management activities is typically performed for vendor audits where the auditee has contracted with the audit sponsor to provide one or more PV services. My primary goal with reviewing this topic is to ensure there is frequent and ongoing oversight of the project, controls and reviews are in place and that there is ongoing communication with the sponsor. I will usually discuss how team members are selected for a specific sponsors project, how training is managed and documented for these team members, the process for the transitioning of employees off and on to the project, the cadence of and topics for meetings with the sponsor and whether there are documented agenda and meeting minutes, how new and updated contracts are managed so the needed project activity changes are disseminated to the project team, ongoing or periodic sponsor deliverables (e.g., weekly or monthly status reports and Quarterly Business Review meetings), and project-specific metrics, including key performance indicators (KPIs) and service level agreement (SLAs).

- Adverse Event (AE)/Product Quality Complaint (PQC) Management

AE/PQC review is the central focus of most PV audits wherever case processing activities are in any form, either full case processing or intake through a call center and submission to the sponsor are contracted. As a result, this topic can be complex or rather abbreviated. Regardless of the scope of activity, the auditee is still responsible for ensuring the services are being performed as contracted. Being a process-oriented person, I usually approach this topic from a logical start to finish process approach. The auditor can approach this session as he/she sees fit but for me it is more logical for me to approach it as a process.

AE/PQC Process

I will begin this session by informing the SME (s) that the procedure(s) have been reviewed but that I would like them to walk me through the process that so we can discuss it in detail. I have not reviewed (or written) any PV procedure yet that captures every detail associated with PV AE/PQC management. This would be challenging as this activity can be nuanced and requires interpretation and analysis of data at various times. Discussion of the process provides me the opportunity to understand the daily operational management of this information.

To begin, I need to understand the predefined workflow for these cases and the sources by the which the auditee will receive this AE/PQC information. AE/PQC sources can be numerous and can include medical information call/contact centers, PAPs, PSPs, HCPs, consumers, family members, social media postings, sponsor websites, sponsor sales representatives, specialty pharmacies, product HUBs, and postmarketed studies to name a few. The method of receipt of this information, where it is documented, the type of system where it is retained, whether the system is 21 CFR Part 11 compliant, and the methodology of the intake, data entry, and processing of the case will be discussed. Auditee roles and responsibilities for each step is reviewed and the method for the storage of AE/PQC source documentation as well as follow up on each case (if contracted) including the tracking of follow ups will be reviewed.

Medical Review

In some instances, sponsors may require the auditee to perform medical review (MR), usually in the situation for CRO vendors performing case processing activities. I will discuss who is performing the medical review, how the case is provided to the medical reviewer (some companies will route through the safety database others will send via email), and controls in place to ensure data integrity and security. The process for conducting medical review will be discussed with the appropriate SME including what data points are reviewed, which data points MR adds to the case, and how the AE causality and listedness ar determined. In some instances, the sponsor will perform the MR internally. The auditee may also be contracted to perform a cumulative medical review of AEs on behalf of the sponsor. This process will also be covered.

Combined AE/PQC Cases

Combined AE/PQC cases are very common in the United States and in the PV postmarketing field in general. Commonly AEs and PQCs are managed separately as there are different FDA regulations that need to be followed. Medical Information Call Center systems have the ability to enter both AE and PQC information as reported, but at the time of sending to the sponsor, they are usually sent to different destinations as instructed by the sponsor. AEs are usually sent to the sponsor's PV department (or vendor), but PQCs are sent to the sponsor's quality department for assessment, management, and follow-up. Commonly, there should be an exchange of case numbers between both parties to cross-reference.

AE/PQC Case Review

Confirmation of the processes reviewed and discussed during the audit are performed during a review of selected AEs/PQCs in scope for the audit. By now, the auditor and auditee have discussed the management of AEs/PQCs in the computer system used for documenting and managing these cases and the procedures used as a baseline for those activities. A review of a sample of cases can confirm the documented processes are being followed. The sample and method by which this occurs is up to the auditor. I usually try to select a sample of different case types processed at various timepoints. In most instances, the auditees will conduct this review within their database while they walk me through the case. This approach allows them to control access to and management of the system while allowing me to view the case information needed to determine if the auditee is compliant with existing procedures, contractual obligations, and regulatory requirements.

Quality Control

Quality control (QC) of AE/PQC case processing is an important part of the overall management of AE/PQC cases. This activity ensures that information received is complete, information entered into the system is accurate compared with the source information provided, data entry errors are corrected and follow up is performed as needed. How and where AE/PQCs cases are submitted, the method for this activity, requirement and receipt of acknowledgement, and tracking will be reviewed. If these activities are not occurring as expected, then I would attempt to understand why this is occurring. QC can occur either as part of the individual case workflow (in line QC), as part of a daily case review or periodically depending on services provided by the auditee. Call center activities may have a process for the review of recorded calls. This process can also be considered as part of the overall QC management for AEs/PQCs.

Translations

Given the global nature of PV activities, safety information can be provided in various local languages. For US-based companies, English is the primary language, so translation of this safety information within an appropriate period of time is crucial. If case processing activities are outsourced the process and timing by which the vendor identifies the need for translation is paramount. A common example is a non-English speaking reporter calls into a contact center, but the call center associate does not speak the callers language. How does the associate manage this call? If the associate does not speak the caller's language, is there a method by which immediate translations services can be utilized? The auditor needs to understand how this scenario may be addressed. Some companies have internal staff who are multilingual and can provide assistance while others utilize translation services where interpreters can join the call to provide translation services. In situations where safety information is provided in written form, typically an AE/PQC form, these documents also need to be translated. Again, there may be internal resources who can perform the translations, or these documents may need to be sent externally to a translation vendor. Ultimately, timing is of the essence as the individuals responsible for the processing, medical review, and analysis of the case need to be able to read the case in the country's primary language (English in the United States) to determine reportability. The auditor will need to determine if there are auditee processes and procedures in place (or perhaps the sponsors procedures are being followed) that outline how these scenarios will be managed as well as the timeline for translation turnaround.

As previously mentioned, many US-based companies have vendors and business/license partner in many countries around the world. Therefore, English may not be the primary language and as a result, the policies and procedures may not be in English. This can be a conundrum for some auditors as often they don't realize this is an issue until the auditee responds to preaudit requests informing the auditor that none of their documents is in English. There are a few ways this can be addressed but the auditor will need to discuss with the sponsor to see if either solution poses a compliance risk for their organization.

One approach is for the English-speaking auditor to utilize a native speaker at the auditee location who can verbally translate the documents for the auditor. This is extremely time-consuming and takes away from audit discussions. In addition, the auditor cannot verify anything the translator is saying which poses a compliance risk to the both the auditor and sponsor. Audits are based on objective evidence, but if the auditor cannot read the procedures and evidence presented, no verification is occurring.

Another approach is that the auditor (in collaboration with the sponsor) can request all relevant audit information and documentation be formally translated by the auditee into English prior to the audit. This can also be time-consuming and costly for the auditee. Some auditees are resistant to this. In some instances, the sponsor can provide assistance in having the documents translated.

I have seen where some auditors utilize online translation platforms for procedures during the audit, but they are limiting in that there is no verification that the translation is correct, and this can also be time consuming for the auditor and take away from the valuable discussions with the auditee.

Ultimately, the auditor can avoid this situation by asking these questions during the planning meeting with the sponsor. The sponsor may not have an immediate answer but ultimately you want the audit to succeed. There may be a simple solution, or the sponsor may need to consider an auditor who is a native speaker.

- AE/PQC Reconciliation

The reconciliation process for AEs/PQCs is a common topic of review associated with PV audits. Such reconciliations can span the simple to the complex. From an PV audit perspective, the number of stakeholders involved in the reconciliation process, including which party initiates the process; how the activity is performed and documented, including the resolution of discrepancies; and method by which the reconciliation are shared and retained are important to understand. It should be noted that reconciliation is not a QC activity although some vendors would disagree with that premise. Quality control of a case and all cases is more detailed than reconciliation. Reconciliation merely provides confirmation that the safety case, for a predefined timeframe, was sent to the appropriate party. The frequency of reconciliation is usually determined by the sponsor, but the initiation and management of activities may be managed by a vendor or business/license partner.

Reconciliation may occur between two parties or there can be one party (e.g., call center) initiating and managing the process as well as performing the reconciliation with various parties (e.g., CRO, SP, PAPs, PSPs, business/license partners, and other call centers). The frequency of these activities vary and can be dependent on the volume of cases, regulatory requirements, and sponsor or vendor procedures. For clinical trials, SAE reconciliation can be managed by the trials data management department (or PV at some companies) and performed at predefined timelines associated with patient enrollment, etc. as outlined in the DMP. Postmarketing AE/PQC reconciliation typically occurs weekly but can occur monthly or quarterly for companies with very low case volume (e.g., 1–4 cases a month). Common data points to be reconciled for companies with marketed products in the United States include the sponsor company name, case number, date/time case was created, case status (open, closed), source of case, contact company, country, state, reporter type, product name, receipt date, and cross link case numbers. See appendix 12.4 for examples.

- Overview of Computer Systems

Computer systems utilized for the intake, storage, and transmissions of safety information are evaluated against FDA 21 CFR Part 11: Electronic Records; Electronic Signatures. Typically, the topics I cover in this section include (but are not limited to) general system information (names, versions, etc.), system location (web based, physical servers, etc.), third-party vendors utilized (including delineation of roles and responsibilities), 21 CFR Part 11 Compliance (including securing and operational controls, audit train capability, validation activities), data protection management, system access, security, maintenance, backups, change control process, business continuity, and evidence of most recent testing and disaster recovery process. Depending on the auditee, the auditor should also consider any applications used for the purposes of managing safety information.

- PV Quality Management System (QMS)

The review of the PV QMS is designed to determine how the auditee is providing oversight and analysis of the various PV activities on an ongoing basis. During the course of the audit, PV procedures and activities would have been discussed, procedures reviewed, and evidence provided. The PV QMS should illustrate how all the individual elements are working together to provide a compliant and effective PV QMS. Oversight of the PV QMS can be conducted in various ways but, commonly, I see various metrics measured to quantify effectiveness of various PV activities. This can include case processing times, submissions times, aggregate report submissions, etc. The auditor should be aware that smaller PV organizations will likely not have a defined PV QMS as they be focusing on building processes and infrastructure for individual processes. Larger organizations may have the ability to monitor and manage the PV QMS on routine basis. Evaluation of the PV QMS is only useful, in my opinion, when the individual PV elements are already in place. Only then can you assess and analyze how the separate parts of the PV systems are functioning together as a cohesive and effective system.

6.4.4 Communication Techniques

Audit communications should be clear, consistent, and frequent between the appropriate parties (e.g. the auditor and sponsor or the auditor and auditee) throughout the various audit stages. The Lead Auditor is responsible for developing and maintaining effective, clear, and consistent communication at all times during the audit process. Communication can aid in the successful execution of the audit, but miscommunication can result in unanticipated issues. It is true that everyone has their own way of listening, interpreting what is being said, and communicating (verbal and nonverbal). The auditor needs to assess the sponsor's and auditees' ways of communicating to ensure appropriate information and responses are being given and received as is intended. This can be difficult at times, but the auditor needs to try to understand if the information being conveyed, and questions asked are resonating and being interpreted correctly by the recipient. The method and techniques by which the auditor addresses these situations can have a significant impact on a positive outcome. Audits require a high degree of interaction if they are to be successful and the auditor needs to be able to obtain all the necessary information to assess compliance with the requirements.

Audit success can depend on the auditor asking the best question at the right time and formulating the questions in such a way as to obtain the desired answer. I will typically start with more general questions and while the conversation evolves ask more specific and targeted questions. There are numerous resources available that describe various communication techniques to be used in specific situations. I will not attempt to discuss all of these techniques here; however, I will discuss a few basic communication techniques I frequently use which have been successful for me.

Open-Ended Questions

The type of question and the way in which an auditor asks a question will hopefully illicit the needed information. However, there is no guarantee this will occur. The auditor may ask a clear and concise question, but the recipient may not understand the question due to a misinterpretation of the question asked, a lack of knowledge regarding the subject matter being discussed or inattentiveness. During an audit, I always try to make the majority of my questions open-ended. These types of questions are designed to allow the recipient the opportunity to provide a response more detailed than a simple Yes or No answer. Open-ended questions usually start with words such as "Where," "When," "What," and "Why.". Phrases such as "Please describe your process for managing AE/PQCs" or "How does your company ensure that all expedited reports have been reported on time by your vendor ?" encourage the auditee to elaborate and provide a detailed response. Open-ended questions are a good way to begin a conversation, illicit information and provide the auditor opportunity to ask more specific questions.

Closed-Ended Questions

Closed-ended questions are questions that will encourage the auditee to provide either a "Yes" or "No" answer. Such questions do not encourage the auditee to provide additional information, but rather, only respond in a "Yes" or "No" manner to the auditor. Examples of such questions can be "Do you have a validated safety database ?", "Are all PV department staff trained on AE/PQC management ?", etc. I always advise against using such questions during audit interview sessions as these questions can shut down a discussion very quickly. The auditor needs the auditee to talk and provide details, but the auditor also needs to ensure the response does not go on so long that the interviewee goes off-topic.

Paraphrasing
I also use a communication technique commonly referred to as paraphrasing (summarizing). This technique is often best used when discussing processes where the auditee will begin a conversation by describing a process (e.g., AE case processing). Once the auditee has completed their review and I have asked some targeted questions for clarity, I will paraphrase the information as it was provided to me. This provides the auditee with the confidence of knowing that their response has been heard and interpreted correctly by me while also preventing me from misinterpreting what I heard from the auditee. I have found this to be immensely successful as I obtained the information needed while continuing to develop rapport and trust with the auditee.

Nonverbal Communication
The auditor should be astutely aware of nonverbal behavior as it relates to auditee' responses to questions asked. This can be a little challenging during remote audits but not impossible. It has been my experience that nonverbal communication can be conveyed not just in body positioning or facial expressions but also in voice tone. In fact, someone's voice tone, for me, is one of the primary indications of how an interviewee may be reacting to questions asked. Everyone has unique personalities and behaves differently in an audit situation. Some interviewees are calm, poised, and communicative; some are arrogant, evasive, abrupt, and defensive; while others appear nervous and overwhelmed as they stumble over their words, provide incomplete sentences, and provide responses that do not answer the question asked. This is really a response to the situation where some perceive an audit as a normal part of doing business in the pharmaceutical/biotechnology industry (which it is), others perceive it as intrusive and inconvenient, while for others this may be their first audit where there is anxiety and fear of making a mistake.

We have all been there, whether working in industry and being the auditee or as the auditor. We all make mistakes so the auditor should be able to interpret the nonverbal behavior of the participants and attempt to engage, reengage, and put the participants at ease. After all, everyone is there to achieve the same goal. As an auditor, I am not seeking to penalize anyone; just gather information in a collaborative and professional manner.

6.4.5 Conflict Resolution

The Cambridge Dictionary (2024) defines conflict as "an active disagreement between people with opposing opinions or principles." The resolution of the conflict (conflict resolution) are the ways and tools used to resolve the conflict so that both sides reach an acceptable agreement. During an audit, conflict is usually the result of some level of miscommunication, misunderstanding, lack of knowledge regarding the topic being discussed, or inattentiveness to the discussion at hand.

Every auditor wants to circumvent conflict during an audit but sometimes this cannot be avoided. Causes of conflict are varied but most commonly are the result of misunderstandings resulting from poor communication, bias, auditee intimidation by the audit process and poor listening. This can be manifested in auditee behavior where there is an identified uncooperativeness resulting from a resistance to the audit being performed, evasiveness to questions, provision of incorrect evidence, an absence of the appropriate SMEs needed to answer the questions, comments about how the audit is not needed and showing a lack of interest in the audit process by not starting sessions on time as evidenced by having other conversations while the audit discussion is being held, performing other work not related to the audit, and not responding to questions asked. If responses to questions are not in line with the question asked the auditor, the auditor can ask the question again in a more concise manner and ask the auditee if the question was understood.

Despite these attempts, the auditor should observe for signals that the auditee is behaving in an uncooperative manner or does not act in a transparent manner. Such behavior can result from the auditee not being engaged in the audit planning process but not reviewing, commenting, or asking questions on the audit plan, audit agenda, or preaudit document request list in advance of the audit. They may have questions but were unwilling to ask these questions ahead of time due to a lack of interest, or irritation in the fact that audit was needed. These can result in the auditee being belligerent, uncooperative, and acting in a manner that jeopardizes the success of the audit. When this occurs, trust between the auditor and auditee is lost.

The auditor needs to manage conflict using the experience, communication skills, training, and techniques at his or her disposal. Initially, once conflict is noted, the best approach is to attempt to diffuse the situation by temporarily stopping the audit. Taking a break can allow all parties to calm down and prevent the auditor from losing control of the audit. Once control of the audit is lost this may result in an abrupt termination of the audit. This is a lose-lose situation for both the auditee and the auditor. Once the auditee calms down the auditor can choose to move on to another topic but plan to revisit the discussion that resulted in the auditee's negative response or have a one-on-one discussion with the auditee. In order to resolve the conflict both parties must listen to the others' perspective on the situation otherwise resolution is not possible. In my experience, I have seen the auditee behave in an unprofessional and belligerent manner as means of trying to intimidate the auditor. Stand strong as the auditor is the director and manager of the audit, not the auditee. In reality, the auditee's unprovoked response is commonly a reaction to questions being asked of them which they consider exploratory and intrusive, which may result in showing a weakness or deficiency. It is important to understand why such a negative response is occurring. In very rare situations, where the auditor considers that auditee response to be hostile to the point in which physical safety is possibly jeopardized it is best to remove yourself from the situation and contact the sponsor to provide an update.

Below are some common principles/techniques I have used to manage audit conflict:

- Actively listen and understand where the other person is coming from.
- Apply constructive feedback techniques.
- Both parties need to agree on the issue.
- Both parties need to agree with the objective.
- Communicate clearly and respectfully.
- Cool down things by temporarily stopping the audit.
- Deal with issues rather than personalities.
- Describe the conflict in objective terms.
- Explore options together.
- Follow the audit standard and guidelines.
- Maintain professionalism and respect.
- Manage your emotions.
- Seek help if needed.
- Understand the other party's perspective.
- Use appropriate conflict resolution styles.
- Use Negotiating and compromising techniques.

6.4.6 Close-Out Meeting

As noted by APIC (2016), "The closing meeting is particularly important since it allows the auditor (or audit team) to communicate the audit findings and conclusions in a logical and coordinated manner to the auditee's management."

In my experience in the pharmaceutical/biotechnology industry and as an auditor, I approach the closing meeting as an opportunity to summarize observations identified during the course of the audit. The closing meeting is not the time where the auditor should surprise the auditee with a list of observations that were never discussed. All observations should be deliberated during the audit sessions and not left to the closing meeting where the auditee is surprised at many of the findings. During the audit opening meeting session, I will inform the auditee that my approach is to work with them in a collaborative manner. That includes informing them of any areas of concerns. These will be pointed out when identified and not left to the closing meeting. As a result, there are no surprises during the closing meeting.

I think this is the most appropriate approach. While working in industry, I have participated in audits where observations were only mentioned during the closing meeting. There was no indication of these observations being an issue during the audit discussions. As result, there was no opportunity to discuss them in a meaningful way or provide additional evidence, if the auditor had requested, that could have avoided the observations to begin with. This puts the auditee in a difficult position and weakens trust between the auditor and auditee.

The closing meeting is usually facilitated by the lead auditor and usually lasts no more than 30 minutes. Remember, at this point everyone is mentally exhausted, and this meeting is meant to summarize identified observations and next steps in the process, so the meeting needs to be clear, concise, and professional. Attendance at the closing meeting is usually the same individuals who have attended the opening meeting but, in some instances, the sponsor representatives may want to attend or there may be additional auditee management attendees.

The way in which the closing meeting is conducted can vary. If the auditor is performing an audit on behalf of a sponsor the sponsor may have a process they want the auditor to follow. This can include the use of PowerPoint template, the inclusion of specific content to discuss including timelines for the provision of the audit report and CAPA plan to the auditee. In some instances, the auditor can use their own PowerPoint presentation, or the closing meeting may be informal where the observations are provided to the auditee verbally. All methods can be equally effective. The key is that the message provided to the auditee is clear and concise. The auditor should be sure to start the closing meeting on time to ensure the appropriate topics are discussed and there is time for the auditees to ask questions.

Some common closing meeting topics that can be discussed include:

- Give an expression of thanks for their participation and work in hosting the audit.
- Remind them of the reason and scope for the audit.
- Remind the auditees that the audit is confidential.
- Present a logical and factual explanation of the strengths and weaknesses of the auditee system.
- Mention that your findings are based upon a sample and that other issues may still exist where you're sampling that you're sampling did not uncover.
- Inform the auditee that any observations discussed during the close out meeting will not be categorized as critical, major, or minor pending the development of the audit report and discussions with the audit sponsor.
- Remind the auditee that the closing meeting will not contain any surprises as all observations were identified and discussed during audit sessions.
- Present the observations clearly and concisely. Additional details will be provided in the audit report.
- Give any statement of recommendation if you are permitted to do so.
- Invite questions and discussions.
- Confirm the auditee primary point of contact for the receipt of the audit report and CAPA plan template (depends on the sponsors audit process).
- Agree to timelines for the provision of the audit report and CAPA plan template.
- Agree to timelines for the auditees provision of the CAPA plan.
- If the auditor will not be the auditee's primary point of contact following the audit, inform the auditee of who will be following up with them from the sponsor.
- Conclude on a positive note.
- Close the meeting.

References

Active Pharmaceutical Ingredients Committee (APIC) (2016) Auditing Guide chrome-extension://efaidnbmnnnibpcajpcglclefindmkaj/https://apic.cefic.org/wp-content/uploads/2021/10/APIC_CEFIC_AuditingGuideAugust2016.pdf
Russell JP (eds) (2013) ASQ Auditing Handbook, Quality Press Milwaukee, Wisconsin
The Cambridge Dictionary (2024) https://dictionary.cambridge.org/dictionary/english/conflict
The World Atlas (2024) https://www.worldatlas.com/articles/how-many-time-zones-are-in-the-united-states.html

Chapter 7
Developing the Audit Report

The development of the audit report is the culmination of all of your hard work during the audit. The approach to the audit report development varies based on the sponsor and can range from a full narrative of each topic discussed during audit, narrative of targeted sections, or an abbreviated format that includes an executive summary and the identified observations. It is important that the auditor remember who the audit report audience will be. In many instances, the sponsor will have the internal Pharmacovigilance (PV) team subject matter experts (SMEs) review the audit report, but in some instances, the audit report will be reviewed by non-PV individuals. In either instance, it is vitally important that the report is concise, clear, and well organized. In this chapter, I discuss some of the major activities associated with the development of the audit report.

7.1 Report Format

The determination of the report format will dictate what information can be added to the report as well as the detail associated with each topic. The audit sponsor may request their audit report template be used or, in some instances, will defer to the auditor to use an acceptable format. It is important to discuss with the sponsor during the planning phase of the audit the requirements associated with the audit report.

There are numerous PV audit report templates that can be used, but the content sections should reflect those topics previously outlined in the agenda. There are also other discussions or topics (not previously in the agenda) that may occur during the audit which are important to add to the audit report. Some common PV audit report sections include, but are not limited, to the following:

- Audit dates
- Auditee point of contact, job title, email address, and telephone number

H. Murphy, *A Guide to Pharmacovigilance Audits in the USA*,
https://doi.org/10.1007/978-3-031-73596-7_7

- Audit location (remote or the physical address of where the audit occurred)
- Auditor name and job title
- Introduction
- Purpose and objective
- Executive summary
- Auditee company background
- Opening meeting
- Audit participants (name and job title)
- Closing meeting
- Quality management system
  - Quality manual
  - Controlled document management
  - Training program (new hire, annual, and sponsor-specific training)
  - Deviation and CAPA management
  - Vendor qualification
  - Audit program
  - Record retention
- Project management activities
- Medical information call center
- AE/PQC management
- AE/PQC cases reviewed
- AE/PQC reconciliation
- Literature search and analysis
- Signal detection
- Aggregate reports
- Computer systems and applications
- Specialty pharmacy inventory management, dispensing process, and packing and shipping process
- Audit observations
- Documents reviewed

There are some sponsors who utilize an abbreviated audit report approach in which the report sections would include:

- Audit dates
- Auditee point of contact, job title, email address, and telephone number
- Audit location (remote or the physical address of where the audit occurred)
- Auditor name and job title
- Introduction
- Purpose and objective
- Executive summary
- Background information
- Documents reviewed
- Audit participants
- Audit observations

If the sponsor requests that the auditor use their template, the auditor needs to ensure the template contains the pertinent information the sponsor expects to see. If you are not certain of report expectations, you need to confirm with the sponsor prior to writing the report. This will prevent numerous revisions which can be time-consuming.

7.2 Organizing the Information

The report template has been determined so it is time to organize the information obtained through document review, interviews, and additional evidence requests. Conducting an audit is stressful. It requires that the auditor multitask while conducting analysis concurrently. Some auditors are naturally organized where each question asked is scripted and documented, and responses with associated evidence are clear and organized. Some auditors are not so organized and may rely primarily on memory. In each case, the auditor needs to consider the numerous pieces of information reviewed and extract the most pertinent pieces of data for inclusion in the audit report. There are many tools the auditor can use for organization, but the auditor must determine what works best.

During my audit experience, I have learned that taking some time to organize my notes and determine the best information to include in the report prior to beginning the writing of the report improves the efficiency of the audit report writing process. It can be difficult to go back and forth from the report to the audit information and audit notes and then back to the audit report. In addition, auditees will provide the auditor additional time following the completion of the audit to access the electronic document exchange platform. This is a good opportunity for the auditor to review any processes or audit requests to confirm details.

7.3 Writing the Report

The audit template has been decided and the audit information has been organized. Now it is time to begin the writing of the report. Each auditor will approach this process based on what works best for them. There is no incorrect approach. It is important that the auditor remember who will be reading the report and incorporate this into how the report is constructed. For example, a PV audit report may be reviewed by PV SMEs, but it can also be reviewed by non-PV personnel. The PV industry uses a lot of abbreviations to represent various activities within PV. Do not assume that these abbreviations are understood by your readers even if you are utilizing a sponsor audit report template. The auditor can always add an abbreviation table which contains the abbreviation and the expanded term or upon the first use of the abbreviation, expand upon the term within the report text. Be clear, concise, and do not make any assumptions. Below is an outline of how I approach the development of a common PV audit reports. This is just my approach based on what works for me.

The first section in any PV audit report is usually the executive summary, which is intended to be brief and to the point. The auditor should remember that this section is intended to provide senior leadership with information on the success of the audit and the issues identified. It should not be more than one (1) page in length. The specific details can be reserved for the individual report sections and observation portion of the report. You can begin with a brief introductory statement of the audit which includes the dates performed, type of audit, and who performed the audit. This is usually followed by the observations identified during the audit. Try to list the observations based on categorization (critical, major, and minor) and provide a brief summary of the issue (one to two sentences). Confirm with the sponsor if they want minor observations listed in the executive summary as some only want the critical and major observations listed in this section. This is typically followed by a conclusions section which highlights the positives of the audit (e.g., vendor is compliant with contractual obligations, etc.) and follows with a high-level summary of issues that need to be addressed.

The executive summary is then followed by the main body of the report. This section of the report contains information on the topics outlined in the audit agenda as well as any additional sections that would be important for the reader to know. The auditor should remember that the audit report is not meant to be a repeat of the auditee's policies or procedures but should give the reader enough information to understand how activities are conducted, confirm controls, and outline identified observations. The report sections need to be approached as summaries of the auditor's evaluation of the topic based on documents reviewed, interviews conducted, and evidence presented.

My approach to the main body of the report is to follow the logical flow of the audit agenda. If some audit sessions are moved to accommodate SMEs, I would suggest the auditor develop the report in a way that works best for communicating the intended message and writing style. I usually begin with an overview of the auditee's quality management system (QMS) which includes the management of controlled documents (system used, new document creation, document revision, and obsoletion), the training program (new hire, annual refresher training, and client-specific training management), deviation and CAPA msanagement, vendor selection, and the audit program. The QMS, in my opinion, provides the foundation and standards for which activities are to be conducted. Following completion of the QMS, subsequent sections can include a review of CAPAs from the previous audit (some sponsors require this) as well as new deviations and CAPAs since the previous audit. Additional sections may include medical call center activities, project management activities, AE/PQC management, signal detection, literature search and analysis, aggregate reports, and an understanding of the computer systems/applications used in support of the project. The computer system discussion may be high level or more specific based on the requirement of the sponsor.

Most audit report templates require a listing of all documents and evidence reviewed prior to and during the audit. I will usually try to provide the listing in the following format:

- Document ID number
- Document name
- Document version
- Document effective date

Upon completion of the section summaries, it is then time to develop the observations.

7.4 Developing Observations

The format, content, and structure of the audit report observations, like many of the other sections of the audit report, will vary based on the audit report template used. Typically, each observation will require a category and subcategory (e.g., case processing/late reporting) and risk level assigned for each observation. Each observation should be clearly and succinctly stated with an objective assessment provided to support the observation. If the observation in the audit report is not presented appropriately and consistently throughout the report, the auditee may not be able to develop an appropriate CAPA plan needed to resolve the issue and prevent reoccurrence. In addition, the reference standard (e.g., regulation, guidance, procedure, contract etc.) should be assigned to indicate to the auditee the reference standard used for the identification of the observation.

The risk categorization for PV audit observations is assessed by the auditor as critical, major, or minor. Depending on the sponsor's requirements, the auditor may also have the opportunity to provide recommendations for improvement. Recommendations are noted areas for improvement identified by the auditor that do not meet the definition for an observation and do not typically require an auditee response. The assignment of observation severity is somewhat subjective based on the auditor's interpretation of the provided definitions, audit information, and evidence. Regardless, audit observations must be fact based and unbiased.

Common postmarked PV inspection observations definitions can include the following:

Critical: Critical observations are serious identified deficiencies in pharmacovigilance systems, processes, or practices that can negatively affect the rights, safety, or well-being of consumers, can pose a public health risk, or represent a significant violation of applicable legislation and guidelines.

Major: Major observations are deficiencies in pharmacovigilance systems, practices, or processes that could potentially adversely affect the rights, safety, or well-being of patients or that could potentially pose a risk to public health or that represents a violation of applicable legislation and guidelines.

Minor: Minor observations are deficiencies in pharmacovigilance systems, practices, or processes that are not expected to have a negative impact on the rights, safety, or well-being of consumers or pose a public health risk.

7.5 Communication of the Audit Report

Communication with the sponsor or auditee during the writing of the audit report is, in my experience, minimal unless the auditor has a report template or very specific clarification questions for the auditee. The audit sponsor will want to review the draft audit report and provide their feedback, questions, etc. In my experience, most of the comments are usually clarification questions from non-PV sponsor personnel requesting additional information on a specific report section. The auditor should carefully review and thoughtfully respond to the sponsor's comments. As we mentioned previously, some of the report reviewers may not be PV personnel and, in many cases, no one from the sponsor attended the audit so they do not have the context of discussions conducted. In addition, the sponsor may have certain elements of the report they want to ensure are documented clearly for internal purposes. There may be several iterations of the audit report depending on the number of comments provided. The sponsor may have questions associated with observation categorization. If a sponsor requests that the auditor change the categorization of an observation, the auditor should use their judgment in making this change. The initial assessment was made based on risk categorization definitions, audit interviews, document and evidence, and the auditor's PV expertise. It is important to have a discussion with the sponsor in this situation. In many cases, sponsor questions are usually based on the need for additional information or clarification from the auditor.

Once all of the sponsor comments are addressed by the auditor, the audit report can be finalized. The audit sponsor may take the lead on the completion of final signatures, or they may defer to the auditor. In most cases, the sponsor will use an FDA 21CFR Part 11 compliant system for electronic signatures. There are some smaller companies that do have such electronic systems and will require the auditor to sign the document, scan it, and send it to the sponsor for finalization.

Once the audit report is signed, the report may or may not be sent to the auditee. This determination is based on the sponsor's audit program procedures. Some companies will send the signed version of the report and a CAPA plan (to address observations) to the auditee while others will send an external summary report to the auditee containing only the audit observations and a CAPA plan. The auditor should confirm with the sponsor how this is to be managed. Some sponsors will want the auditor to send the report and CAPA plan to the auditee, while other sponsors will manage this internally.

Chapter 8
CAPA Plans and Audit Closure

In this chapter, we will discuss generally the role of the PV auditor in the review and approval of auditee Corrective and Preventative Action (CAPA) plans following the completion of the audit as well as activities associated with the closure of the audit.

8.1 CAPA Plans

Corrective and Preventative Action (CAPA) plans can be required to be completed by the auditee to address audit observations. The CAPA plan is intended for the auditee to outline the root cause analysis, corrective actions, and preventative actions for each observation. Anticipated completion dates and the contact for ownership of the CAPA are also to be documented on the CAPA plan. The CAPA plan template is usually provided by the audit sponsor to the auditee with a date for submission of the CAPA plan for review. The CAPA plan is separate from the tracking of individual CAPAs and provides an overview of the intended activities for completion by the auditee.

The sponsor may request the auditor to participate in the review of the CAPA plan to ensure the auditee is adequately addressing the issues that lead to the assignment of the observation by the auditor. The auditor has the cumulative knowledge based on documents reviewed, interview sessions, evidence provided, and knowledge of the associated regulations. For some sponsors, a review of the CAPA plan by the auditor ensures the auditee is adequately addressing the deficiency that resulted in the observation.

During the process of reviewing the CAPA plan, the auditor can accept the plan, reject the plan or specific elements of the plan, or request additional clarification. The auditee must ensure that the proposed CAPA plan identifies the root cause of each deficiency, provides an immediate corrective measure, and develops and implements preventative measures to prevent a reoccurrence of the issue. The

H. Murphy, *A Guide to Pharmacovigilance Audits in the USA*,
https://doi.org/10.1007/978-3-031-73596-7_8

auditee must not simply perform a corrective action with no root cause analysis or preventative actions. The auditor reviewing a CAPA plan should not readily accept human error as a root cause. There are often reasons for human errors to occur, including inadequate employee training, incorrect assignment of an employee to perform a specific job, insufficient employee oversight and management, unclear procedures, and additional resource issues. Human error is the result of one of these deficiencies and not necessarily the causal factor(s).

Individual CAPAs are normally tracked by the sponsor to ensure completion timelines align with internal procedures. The timelines associated with individual CAPAs can occur over a period of months for more extensive remediation, while others can be quickly remediated. The sponsor may additionally request that the auditor review individual CAPA evidence (e.g., new SOPs, Work Instructions (WIs), computer system validation documentation, etc.) to ensure the auditee remediation activities effectively address the observations.

8.2 Audit Closure

The audit sponsor will determine when the audit is considered to be closed based on internal procedures. In general, the audit is usually considered as closed following the provision of an audit certificate from the auditor and upon completion of all auditee CAPAs. Closure of each CAPA is based on the evidence provided associated with corrective and preventative actions. Each sponsor has their own requirements for audit closure so the auditor should ensure the sponsor's requirements for this activity are clarified.

Chapter 9
The Do's and Don'ts of a Successful Audit

Below is a summary of the various concepts and auditor tips discussed in this book as well as other considerations that can impact the success of the Pharmacovigilance (PV) audit. Some of these concepts are self-explanatory but are still worth mentioning. The auditor should keep these tips in mind during all phases of the audit process.

Do's	Don'ts
Plan for the audit (research auditee, read contracts, preaudit request documents, etc.)	Appear unprepared or disorganized
Hold a preaudit meeting with the audit sponsor to clarify the audit purpose and scope	Argue with the auditee or sponsor
Clarify document templates to be used (e.g., sponsor or auditor)	Make assumptions regarding any aspect of the audit. Always clarify with the sponsor or auditee
Ask the sponsor if there are auditee issues that need to be explored during the audit	Attempt to assume a position of authority
Conduct yourself in a professional manner at all times whether conducting a remote or on-site audit (appearance, communication, etc.)	Discuss personal issues during the audit
Communicate clearly with the sponsor and auditee (written and verbal)	Interrupt anyone when they are trying to answer your question
Seek support from the sponsor if the auditee is being uncooperative or abusive	Allow the auditee to direct the conversation
Take a break if you think the audit discussion is becoming contentious	Allow the auditee to be rude, argumentative, or disrespectful
Remain in control of the audit even if a conflict arises	Be rude at any time
Stay up-to-date on PV best practices, regulations, and guidance	Become distracted or eat while on camera
Discuss an identified noncompliance for a topic not outlined on the agenda with the auditee and sponsor	Ignore an identified noncompliance for a topic not outlined on the agenda
Select audit team members with a holistic approach to include not only audit experience but also communication, topic expertise, etc.	Select audit team members solely based on audit experience criteria

H. Murphy, *A Guide to Pharmacovigilance Audits in the USA*,
https://doi.org/10.1007/978-3-031-73596-7_9

Do's	Don'ts
Be aware of the experience of those attending the audit	Impose the auditor's time zone when conducting an auditor with global team members
Understand how an auditee noncompliance can impact the audit sponsor's compliance	Appear biased in any way
Clearly establish expectations and clear communication with all audit team members	Avoid discussing agenda topics for which the auditee may hold strong opinions or have objections
Be mindful of cultural norms	
Develop audit tools that allow for effective audit planning and audit conduct	

Appendices

Audit Plan Example Templates

Audit Plan (Vendors)

Audit Details			
Audit Number (per sponsor system)			
Title of audit			
Scheduled Start Date		Scheduled End Date	
Audit Type		Lead Auditor	

Audit Scope
[add company name} as a Marketing Authorisation Holder/Product License Holder, must have in place an appropriate pharmacovigilance system to assure responsibility and liability for all of the company products. The supplier/vendor audit program will ensure that both [company and vendor] are compliant with the pharmacovigilance obligations as defined in the terms of the Master Services Agreement and/or separate Statement of Works (SOWs), PVAs/SDEA as applicable, thereby ensuring that all national/international requirements with respect to safety information management are fulfilled. This is a [qualification/routine/for cause] audit of pharmacovigilance activities performed by the [vendor name] to assess the compliance of the pharmacovigilance system with applicable company and/or vendor procedures, with global and local regulations, and to review the interfaces and interdependencies between company and the vendor. The scope of the audit will cover the services by the vendor as it relates to [add summary of services] outlined in the Master Services Agreement and associated Statement of Work (s) (SOWs).

H. Murphy, *A Guide to Pharmacovigilance Audits in the USA*,
https://doi.org/10.1007/978-3-031-73596-7

The audit will encompass review of the services described in the Master Service Agreement and Statement of Work (s) by and between company and vendor, including but not limited to the following: Add a list of topics to be discussed here… Audit samples will be selected on a risk-based approach during the audit in order to review outputs of the processes under review.
Audit Purpose
The audit will be designed to answer the following objective questions: To determine if [vendor] , are in compliance with regulatory and contractual requirements including how issues associated with the process are managed To determine whether deviations and quality issues related to (add scope) processes are adequately reported, investigated, corrected/mitigated, and resolved to avoid future reoccurrences To determine whether contracted services are being performed as planned and reported to company To determine whether personnel training is adequately performed and documented. Confirm whether computer systems used in management of contracted services are compliant as outlined per regulations and their intended use. Add more bullets as needed …
Audit Methodology
• Lead Auditor: add name…on behalf of add sponsor company name • Audit Sponsor: Add sponsor and sponsor vendor leads • Audit Sponsors responsibilities include: • The audit will be conducted from dd-dd mmm yyyy to dd-dd-mmm-yyyy and will be conducted onsite [add address] or remotely via [add meeting application[. • A closing meeting will be scheduled following the end of audit conduct to facilitate a review of the audit findings.

- The audit methods include but are not limited to the following:
  - Pre-audit questionnaire and documentation requests prior to the audit
  - Interviews with staff involved in the process
  - Document Reviews including but not limited to the following:
  - Procedural documents
  - Organizational Charts
  - Deliverables to demonstrate compliance with the MSA/SOWs
  - Business Continuity Plans
  - Disaster Recovery Plan
  - A selection of training records, resumes/CVs, and job descriptions
  - A selection of QC documentation demonstrating compliance with each process
  - A selection of case files
  - Contracts/agreements of outsourced activities where applicable
  - Evidence of computer system compliance/validation as per [21CFRPart 11/Annex11] used to support sponsor activities

Audit Standards

- The audit standards for this audit include but are not limited to:
  - Relevant vendor procedural documents (where applicable)
  - Safety Data Exchange Agreement between company and vendor (add names)
  - Pharmacovigilance Agreement between company and vendor (add names)
  - Applicable Regulatory Requirements including but not limited to:
  - CFR Title 21 Part 11 Electronic Records, Electronic Signatures
  - CFR Title 21 Part 314 Applications for FDA Approval to Market a New Drug (Sec. 314.80 Post marketing reporting of adverse drug experiences)
  - ICH E2D: Post-approval Safety Data Management: Definitions and Standards for Expedited Reporting
- Add additional regulations, guidances…
- Add in additional regulations, standards and procedures as needed….

Audit Plan (Business/License Partner)

Sponsor Name	Add full name	
Auditee Name and Address	Add address and indicate if the audit was remote	
Audit Type	E.g. Routine CSP PV Systems Audit	
Primary Contacts	Add sponsor name and contact information	
	Add auditee name and contact information	
Audit Number		
Audit Dates	DD-MMM-YYYY to DD-MMM-YYYY	

Audit Objectives (Purpose)
The objectives of this [add audit type] audit of [business partner/vendor] are to:

- Review capability, infrastructure, and processes and verify the availability of a Pharmacovigilance (PV) Quality Management System (QMS)
- Review the processes for the identification, collection, evaluation, and submission of Adverse Events (AEs)/Product Quality Complaints to meet local and global regulatory requirements, processes, and company standards
- Determine whether adequate contractual arrangements are in place for the delegation of PV activities and responsibilities;
- To assess the oversight and exchange of information/safety data from sources the MAH has contractual arrangements, e.g., third parties such and marketing partners, license partners, distributors, and other service providers;
- To assess whether contractual obligations are being met
- Verify that key personnel are appropriately qualified, trained, and experienced to perform their assigned PV tasks and that roles/responsibilities are clearly documented through the review of JDs, CVs, and training records
- Review PV computerized/information technology (IT) systems, databases, and applications to assess compliance with regulatory requirements and company policies and procedures
- Review the procedure for the management and archiving of PV information

Audit Scope
In order to determine if [business partner/vendor] PV obligations and responsibilities are being met, the scope of the audit will include (but is not limited to) review the following:

- Determine the MAH's compliance with regulatory PV obligations and responsibilities and provide affirmation that the PV obligations of the auditee/MAH are being met.

- Appropriate organization and resourcing, and have adequate systems and infrastructure in place to perform the relevant PV activities.
- Verify the availability and implementation of a PV QMS, in particular the production, management, and implementation of SOPs, and assure that they follow applicable regulatory requirements and guidelines. Identify areas of noncompliance and methods for remediation.
- Assess the delegation of PV activities and oversight of third-party service providers, business/license partners and distributors by the MAH, and determine whether contractual obligations are being met.
- Determine if PV-related documentation is being adequately controlled, retained, securely stored, and archived as per regulatory requirements, guidelines, and company standards.
- Review training procedures, plans, curriculums, records, CVs, and job descriptions (JDs) for personnel involved in PV activities (including contracted staff).
- Processes and procedures for the identification, collection, processing, management, quality control, medical review, follow-up for missing information, AEs/PQCs arising from any source.
- Timely dissemination and submission of safety data internally within the company and to external business partners, and expedited reporting of Individual Case Safety Reports (ICSRs) to regulatory authorities.
- CAPA management of safety noncompliance.
- Processes and procedures for the management and handling of pregnancy reports and outcomes, including reports of special interest (e.g., medication error, overdose, misuse, abuse, occupational exposure, lack of efficacy, off-label use).
- Reconciliation activities (e.g., receipt verification of safety-related data) between the sponsor and business partner/vendor.
- Review of validation status, testing, security controls, back-ups, change control, user access of computerized/IT systems, databases and applications that support the conduct, management, and handling of PV-related activities.
- Business continuity and disaster recovery plans with evidence from most recent testing.
- Risk-benefit analysis and signal detection/safety surveillance activities.
- Scheduling, preparation (including data evaluation and control), review, approvals, and submission of the aggregate reports Periodic Safety Update Reports (PSURs)/Periodic Benefit-Risk Evaluation Reports (PBRERs)/Periodic Adverse Drug Experience Reports (PADERs), etc.
- Responding to requests and meeting commitments from regulatory authorities.
- Procedures for handling communications to and from HCPs and patients/consumers relating to safety data and management of medical information inquiries and questions.
- Procedures for handling management of safety data arising from sales and marketing activities.
- Process for handling of product quality complaints from spontaneous sources (including those associated with AEs).

- Systems and processes related to drug handling/distribution for postmarketing activities, including product recall procedures.
- Procedures for the management and handling of solicited and unsolicited safety data arising from market research and health economics/outcomes activities (e.g., market research surveys, questionnaires, registries, structured data collection programs, compensation/reimbursement scheme).

Audit Standards

The audit will be conducted to assess compliance with applicable sections of current legislation, directives, and other global regulatory guidelines, including contractual agreements, as follows:

- ICH Periodic Benefit-Risk Evaluation Report (E2C) (R2)
- ICH Post-Approval Safety Data Management: Definitions and Standards for Expedited Reporting (E2D)
- ICH Pharmacovigilance Planning (E2E)
- ICH Pharmaceutical Quality System (Q10)
- 21 CFR 310.305 Postmarketing Safety Reports for Human Drug and Biological Products; Electronic Submission Requirements
- 21 CFR 314.80 Postmarketing Reporting of Adverse Drug Experiences
- 314.81 Other Postmarketing Reports
- 314.98 Postmarketing Reports
- FDA Guidance for Industry – Guidance of the Development and Use of Risk Minimization Actions Plans (RiskMAP) (March 2005)
- FDA Guidance for Industry – Good Pharmacovigilance Practices and Pharmacoepidemiological Assessment (March 2005)

Audit Preparation and Schedule

The auditor will develop a preaudit document request list to obtain initial documentation for review in preparation for the audit. Based on the responses, the auditor may have additional requests.

Audit Conduct

[product name] will be in scope of this audit using a sample-based approach to review AE/PQC case files and aggregate reports and other PV-related processes associated with the sample selection.

The audit conduct will be comprised of interviews with key personnel and document review. An opening meeting will be held on the first day of the audit and a close-out meeting, to provide oral feedback, will be held on the last day of audit. No categorization (critical, major, minor) of observations will be provided at the time of close-out meeting.

An opening meeting will be conducted on Day 1, which will consist of:

- Introductions: Introduce personnel, roles, and responsibilities
- Review of the audit scope and objectives
- Review of the audit agenda to confirm sessions and SME availability
- Logistics: Discuss audit timings, document requests, and other logistics

- [scope summary] The audit will encompass review of the following areas at business partner/vendor:

- AEs/PQCs occurring with [product] and other interventional and noninterventional program and regulatory submissions activities
- Organizational QMS
- PV QMS
- Aggregate report development and submissions
- Medical affairs
- Product quality
- Medical information
- Market research
- IT software/hardware support of PV systems, databases, and applications

Close-Out Meeting

A close-out meeting will be conducted by the lead auditor at the conclusion of the final day of the audit. A review of the audit scope and objectives, including additional areas reviewed, will be summarized. Observations identified during the course of the audit will be discussed and the process following completion of the audit (e.g., Audit report and CAPA plan) will be reviewed.

Audit observations will not be classified during the close-out meeting.

Audit Report

Following the audit, a confidential draft initial audit report will be prepared and issued to [sponsor] within [add number] of business days of the audit.

The audit report will contain a [detailed overview/summary of key findings] of the audit, and present any observations in the form of audit observations with a classification rating and observation category assigned.

CAPA Plans

The auditor will review [Auditee]'s proposed CAPA Plan to ensure observations are being appropriately addressed. [sponsor] will be responsible for the tracking and following of all CAPA plan responses.

Combined Audit Plan and Agenda (General Use)

AUDIT PLAN AND AGENDA

AUDITEE: Add the company name and address

AUDITEE CONTACT: Add the name and address of primary contact

TYPE OF AUDIT:

AUDIT APPROACH: Onsite/Remote

AUDIT PURPOSE: The purpose of this onsite/remote audit is to ……

AUDIT SCOPE: The scope of this routine PV audit is to provide [sponsor name] with a systematic, independent, and objective evaluation of current services provided by [vendor name]. Contracted services include……

AUDIT DATE(S): dd mmm yyyy – dd mmm yyyy

AUDIT STANDARDS: add in applicable local and global regulations vendor SOPs, applicable contracts, SDEAs, PVAs, etc.

AUDITOR(S): Lead Auditor Name

PURPOSE AND SCOPE

This is a routine Pharmacovigilance audit to ensure [vendor name] has the appropriate capabilities to ensure compliance with applicable regional and international PV legislation and guidance between [company name and vendor name].

The audit scope will include a review of [vendor name]'s Quality Management System (QMS), AE/PQC processing and submissions, aggregate report management, safety governance, computer systems/applications utilized in support of [company name] activities, business continuity, and disaster recovery.

Below is the proposed agenda for this [add in number of days] audit.

AUDIT AGENDA	DAY 1 – DD MMM YYYY	
Time (EST)	Interview Topic	SMEs
9:00am-9:15am	Introductions	
9:15am-9:45am	Company Overview	

Additional tables for subsequent audit days can be added. See audit agenda templates for additional information.

Audit Agenda Templates

One-Day Audit Template

Audit Agenda			
Day#/ Dates	Time EST	Session Description:	Attendees (Name/Title):
Day 1/ dd-mmm-yyyy	9:00am - 9:15am	Opening Meeting • Introductions (auditor and auditee) • Review audit objective and scope • Review audit agenda • Audit logistics	
Day 1/ dd-mmm-yyyy	9:15am – 9:30am	Company Overview • History and overview (*please provide a brief presentation*) • Services, sponsors, organizational structure, alliances/partnerships, resources • Regulatory inspection history	
Day 1/ dd-mmm-yyyy	9:30am - 10:15am	Overview of the Quality Document Management System/Learning Management Systems/CAPAs/Audits • Document Management System • LMS- Training System • Review of Training Files for [sponsor] project personnel (*Auditors will select documents for review*) • Deviation and CAPA Management • Internal Audit Program • Vendor selection and oversight	
Day 1/ dd-mmm-yyyy	10:15am - 10:30am	Break	

Audit Agenda			
Day#/ Dates	Time EST	Session Description:	Attendees (Name/Title):
Day 1/ dd-mmm-yyyy	10:30am-11:00am	Project Management Activities • [Sponsor company] project specific training • Contract Management • Project Meetings • Reporting of key metrics to sponsor	
Day 1/ dd-mmm-yyyy	11:00am-11:30am	Review of CAPAs from the yyyy audit & review of [sponsor name] CAPAs since the last audit	
Day 1/ dd-mmm-yyyy	11:30am – 12:00pm	Overview of Computer Systems used for the [sponsor name] Project • General info (Names, versions, etc.) • Location (web based, physical servers, etc.) • Third party vendors utilized (if applicable) • 21 CFR Part 11 Compliance • Data Protection Policy, Patient Confidentiality • System Security, Maintenance, backups • Change control • Business continuity plan & evidence of periodic testing • Disaster recovery process & evidence of recent testing • Data privacy	

Audit Agenda			
Day#/ Dates	Time EST	Session Description:	Attendees (Name/Title):
Day 1/ dd-mmm-yyyy	12:00pm - 1:00pm	Lunch	
Day 1/ dd-mmm-yyyy	1:00pm-2:30pm	AE/PQC Management • AE/PQC Corporate & PV staff [vendor] Training • AE/PQC sources of information • Translation Management (if applicable) • Quality Checks • Medical Review • Case Follow-up • Case closure • Reconciliation • Compliance tracking for submitting ICSRs per contractually defined timelines to [sponsor] • Compliance/KPI monitoring and communication to [sponsor] • Maintenance of case files and source data *Auditor will select AE/PQC cases to review*	
Day 1/ dd-mmm-yyyy	2:30pm – 4:30pm	Document Review	
Day 1/ dd-mmm-yyyy	4:30pm-5:00pm	Audit Close Out Meeting	

Two-Day Audit Template

Day 1: DD MMM YYYY		
Time (US Eastern Time)	Audit Topic	Auditee(s) (initials)
9:00-9:30	Opening Meeting • Introductions (auditor and auditee) • Review audit objective and scope • Review audit agenda • Audit logistics	
9:30-10:00	Company Overview • History and overview (*please provide a brief presentation*) • Services, sponsors, organizational structure, alliances/partnerships, resources • Regulatory inspection history	
10:00-11:00	QMS-Overview of the Quality Document Management System/Learning Management Systems • Quality Manual • Controlled document lifecycle (including eDMS) • Training: Induction/continuous PV training, Interdepartmental training requirements, curriculums, and oversight (including eLMS) • Job descriptions and Curriculum Vitae (CVs) management • CAPA processes, management of nonconformances and deviations, customer escalation process (including electronic system used) • Vendor Management	

Day 1: DD MMM YYYY		
Time (US Eastern Time)	Audit Topic	Auditee(s) (initials)
	• Management of audits and inspections • Documentation, archiving & record management (paper vs. electronic) *Auditor will select JDs, CVs, and training records for review.*	
11:00-11:30	Review of CAPAs from the [add previous years] audit	
11:30-11:45	Break	
11:45-12:30	Medical Information Call Center • Overview of CRM system • Sources of AEs/PQCs • Identification, collection & management of AE/PQC information • Management of Medical Information Inquiries • Record Retention	
12:30-13:30	Lunch	
13:30-14:30	Adverse Event/Product Complaint Management • AE/PQC Training • Sources of information • Translations • Receipt, tracking, duplicate search, source data verification • Medical Review/Causality Assessment • Handling of follow-up information • Invalid Cases • Quality Checks • Combined AE/PQC management • Case Closure	

Day 1: DD MMM YYYY		
Time (US Eastern Time)	Audit Topic	Auditee(s) (initials)
	• Record Retention • Reconciliation with all sources • Submission to Health Authorities • Compliance and KPI monitoring • PV Intelligence *Auditor will select AE/PC cases for review by vendor team in the database*	
14:30-15:00	AE/PQC Case review in the database	
15:00-15:30	Literature Review & Social Media Monitoring • Local Literature Monitoring Process • Identification of AEs/PCs • Social Media monitoring or programs • Communication with [sponsor] • Record Retention	
15:30-16:00	Aggregate Reports • Development and management • Communication and exchange with Business Partner • Regulatory Submissions • Report Tracking • Record Retention	
16:00-16:30	Document Review	Auditor
16:30-17:00	Review of outstanding requests and Day 1 debrief (if needed)	
	End of Day 1	

Day 2: DDMMMYYYY		
Time (US Eastern Time)	Audit Topic	Auditee(s) (initials)
9:00-9:15	• Follow up from Day 1 • Status of outstanding requests	
9:15-9:45	• PV Training • Corporate AE/PQC Training • Vendor Training • Record Retention	
9:45-10:15	PV Vendor Oversight • Methods and frequency of vendor oversight monitoring Overview of process • KPIs/SLAs • Management of deviation from KPIs	
10:15-11:30	Document Review	
11:30-12:30	Overview of Computer Systems • General info (Names, versions, etc.) • Location (web based, physical servers, etc.) • Third party vendors utilized (if applicable) • Validation (21 CFR Part 11) • Data Protection Policy • System Security, Maintenance, backups • Change control management • Business continuity plan/evidence of recent testing • Disaster recovery process & evidence of recent testing	
12:30-13:30	Lunch	

Day 2: DDMMMYYYY		
Time (US Eastern Time)	Audit Topic	Auditee(s) (initials)
13:30-14:00	Signal Detection • Signal validation, investigation, and determination • Reporting of signals to the FDA • Management of Health Authority requests • Communication with [sponsor]	
14:00-14:30	Label Management • Process for determining the need for a label change • Management of Label changes • Review and management of labelling artworks • Record Retention	
14:30-16:30	Document review and preparation for close out	
16:30-17:00	Audit Close Out Meeting	
	End of Audit	

Auditor can add additional tables for subsequent audit days.

CRO Audit Agenda Template: Example 1

Proposed Agenda				
Day	Time (EST)	Agenda Session	Lead SME	Comments
Day 1: DDMMMYYYY	08:00 – 08:30	Opening Meeting • Introductions (auditor and auditee) • Review audit objective and scope • Review audit agenda • Audit logistics		
Day 1: DDMMMYYYY	08:30-09:00	Company Overview • History and overview • Services, sponsors, organizational structure, alliances/partnerships, resources • Subcontracting (third party suppliers) • Regulatory inspection history		Please provide a brief Presentation
Day 1: DDMMMYYYY	09:00-10:00	Quality Management System • Life-cycle management of Procedural Documents, including training and deviations • Monitoring of (local) PV legislation		

Proposed Agenda				
Day	Time (EST)	Agenda Session	Lead SME	Comments
		• Training: Induction/continuous PV training, Interdepartmental training requirements, curriculums, and oversight • Job descriptions and Curriculum Vitae (CVs) • Management of audits and inspections • Handling of incidents, deviations and CAPA management • Inquiries from Regulatory Authorities • Documentation, archiving & record management (paper vs. electronic) *Auditor will select JDs, CVs, and training records for review.*		
Day 1: DDMMMYYYY	10:00-10:15	Break		

Proposed Agenda				
Day	Time (EST)	Agenda Session	Lead SME	Comments
Day 1: DDMMMYYYY	10:15-10:45	Review of [sponsor company] deviations and CAPAs		
Day 1: DDMMMYYYY	10:45-12:00	Overview of Computer Systems • General info (Names, versions, etc.) • Location (web based, physical servers, etc.) • Third party vendors utilized (if applicable) • Validation (21 CFR Part 11)/Annex 11 • Data Protection Policy • System Security, Maintenance, backups • Change control ○ Business continuity plan ○ Disaster recovery process & evidence of recent testing		Please provide a brief presentation
Day 1: DDMMMYYYY	12:00-13:00	Lunch	All	
Day 1: DDMMMYYYY	13:00-14:30	[sponsor program #1 -add title] • Receipt of Safety Information • Query Management • ICSR Submissions • ICSR Tracking		Please provide flowcharts for discussion purposes if available

Proposed Agenda				
Day	Time (EST)	Agenda Session	Lead SME	Comments

		• Aggregate Reports
		  ○ Submissions Regulatory Authorities & Business/License Partners
		  ○ Tracking of Aggregate Report Submission Dates
		• Database Management
		• Filing of ICSRs and Aggregate Report submissions in the TMF
		• Selected ICSR cases for review in the database.
				(auditor will randomly select cases/documentation evidence for review)

Day 1: DDMMMYYYY	14:30-15:30	Document Review	Auditor	
Day 1: DDMMMYYYY	15:30-16:00	Day 1 review of outstanding audit requests and debrief (if needed)		
Day 2: DDMMMYYYY	08:00-08:30	• Review of outstanding requests from Day 1 • Review of audit agenda for Day 2		

Proposed Agenda				
Day	Time (EST)	Agenda Session	Lead SME	Comments

Day 2: DDMMMYYYY	08:30-09:30	[sponsor program #2-add title]
		• Receipt of Safety Information
		• Query Management
		• ICSR Submissions
		  • ICSR Tracking
		• Aggregate Reports
		  • Submissions to
		  • Tracking of Aggregate Report Submission Dates
		• Reporting Requirements for applicable countries (including Japan)
		• Selected ICSR cases for review in the database.		

Day 2: DDMMMYYYY	09:30-10:00	[sponsor program #3-add title] • Preparation of Expedited safety information • Regulatory Submissions		Please provide flowcharts for discussion purposes if available ***(auditor will randomly select cases/documentation evidence for review)***
Day 2: DDMMMYYYY	10:00-10:15	Break		

Proposed Agenda				
Day	Time (EST)	Agenda Session	Lead SME	Comments

Day 2: DDMMMYYYY	10:15-10:45	[sponsor program #4-add title]
		
		• ICSR Processing
		  ○ Intake of Safety Information from ex-North American sites and notification to the Sponsor
		  ○ Query Management for ex-North American Sites
		  ○ Submission of expedited reports
		  ○ ICSR Tracking
		
		• Aggregate Reporting
		  ○ Submissions to regulatory authorities
		  ○ Tracking of Aggregate Report Submission Dates
		  ○ Determining and Management of Reporting Requirements
		• Filing of ICSR and Aggregate Report submissions
				Please provide flowcharts for discussion purposes if available
				
				(auditor will randomly select cases/documentation evidence for review)

Proposed Agenda				
Day	Time (EST)	Agenda Session	Lead SME	Comments
Day 2: DDMMMYYYY	10:45-12:00	[sponsor program #5-add title] • ICSR Processing o Intake of Safety Information and notification to sponsor o Query Management o Submission of ICSRs regulatory authorities o ICSR Submission Date Tracking • Aggregate Reporting • Submissions to regulatory authorities • Tracking of Aggregate Report Submission Dates o Determining and Management of Reporting Requirements • Filing of ICSR and Aggregate Report submissions		Please provide flowcharts for discussion purposes if available ***(auditor will randomly select cases/documentation evidence for review)***
Day 2: DDMMMYYYY	12:00-13:00	Lunch	All	
Day 2: DDMMMYYYY	13:00-14:00	[Sponsor program #6-add title] • ICSR Submissions • Tracking of ICSR		Please provide flowcharts for discussion purposes if available

Proposed Agenda				
Day	Time (EST)	Agenda Session	Lead SME	Comments
		Submission Dates • Aggregate Reporting submissions to regulatory authorities • Tracking of Aggregate Reporting Submission dates • Determining Reporting Requirements • Filing ICSR and Aggregate Report submission dates		***(auditor will randomly select cases/documentation evidence for review)***
Day 2: DDMMMYYYY	14:00-15:30	Document Review & preparation for close out meeting	Auditor	
Day 2: DDMMMYYYY	15:30-16:00	Audit Close Out Meeting		

CRO Audit Agenda Template: Example 2

<table>
<tr><td colspan="3">AGENDA DAY 1 – DD MMM YYYY</td></tr>
<tr><td>Time (EST)</td><td>Topic</td><td>SMEs</td></tr>
<tr><td>9:00am-9:15am</td><td>Opening Meeting
• Introductions (auditor & auditees)
• Review audit objective and scope
• Review audit agenda
• Audit Logistics</td><td></td></tr>
<tr><td>9:15am-9:45am</td><td>Company Overview
• History and overview (please provide a brief presentation)
• Services, organizational structure, alliances/partnerships, resources
• Regulatory Inspection history</td><td></td></tr>
<tr><td>9:45am-10:45am</td><td>Quality Management System
• QMS overview
• Document Control
• Management of audits and inspections
• Deviations and CAPA management
• Documentation, archiving & record management</td><td></td></tr>
<tr><td>11:00am- 12:30pm</td><td>ICSR Management (Expedited and non-expedited cases)
• Initial notification from the clinical sites
  • Safety Mailbox management
  • Initial notification to [sponsor]
  • Case Processing Workflow
  • Case Quality Checks
  • Medical Review & Assessment Process
  • Case Follow Ups
  • Case Translations (if applicable)</td><td></td></tr>
</table>

AGENDA	DAY 1 – DD MMM YYYY	
Time (EST)	Topic	SMEs
	• Case Submissions to Health Authorities • Case Closure Activities • Case Deactivation/Nullification Activities • Reconciliation Process • Maintenance of case files and source data • Compliance and KPI monitoring • ICSR Tracking Reports *Auditor will select cases to be reviewed in the Safety database*	
12:30pm-1:00pm	AE Case review	
1:00pm-1:45pm	Lunch	
1:45pm-3:00pm	AE Case review continued & discussions continued if needed	
3:00pm-4:30pm	Document Review	
4:30pm-5:00pm	Review of outstanding requests and Day #1 debrief if needed	
9:00am-9:30am	Review of outstanding requests from Day #1/Follow up questions	
9:30am-10:00am	Aggregate Reports • Process overview for PSUR development • Process for Line Listing development • Quality Review process • Report Submissions and tracking • Document retention and storage	

AGENDA	DAY 1 – DD MMM YYYY	
Time (EST)	Topic	SMEs
10:00am-10:45am	Literature Search & Review Activities • Process Overview • Search Strategy Management • AE identification & management	
10:45am-11:00am	Break	
11:00am-11:30am	Overview of Computer Systems/Applications in support of the [sponsor] project • General info (Names, versions, etc.) • Location (web based, physical servers, etc.) • Third party vendors utilized (if applicable) • Validation • Data Protection Policy • System Security, Maintenance, backups • Change control ○ Business continuity plan and testing ○ Disaster recovery process and testing	
11:30am-12:30pm	Training • New Employee Onboarding • Annual training • Training curriculum, LMS, etc. • Job descriptions and Curriculum Vitae (CVs)	
12:30pm-1:00pm	Lunch	
1:00pm-4:30pm	Document Review & prepare for close out	
4:30pm-5:00pm	Audit Close out meeting	

Specialty Pharmacy Audit Agenda

Day#/ Dates	Time- US Eastern Time	Session Description:	Attendees (Name/Title for each session)
Day 1/ dd mmm yyyy	8:30am-8:45am	Opening Meeting • Introductions (auditor and auditee) • Review audit objective and scope • Review audit agenda • Audit logistics	
Day 1/ dd mmm yyyy	8:45am-9:15am	Company Overview • History and overview *(please provide a brief presentation)* • Services, sponsors, organizational structure, alliances/partnerships, resources • Subcontracting (third party suppliers) • Regulatory inspection history	
Day 1/ dd mmm yyyy	9:15am-10:15am	Quality Management System • Life-cycle management of Procedural Documents • Training • Deviation and CAPA Management • Job descriptions and Curriculum Vitae (CVs)	

Day#/ Dates	Time- US Eastern Time	Session Description:	Attendees (Name/Title for each session)
		• Management of audits and inspections • Record Retention *Auditor will select JDs, CVs, and training records for review.*	
Day 1/ dd mmm yyyy	10:15am-10:45am	Break	
Day 1/ dd mmm yyyy	10:45am-11:15am	Review of CAPAs since the previous audit (if applicable)	
Day 1/ dd mmm yyyy	11:15am-12:00pm	Overview of Computerized Systems • General info (Names, versions, etc.) • Location (web based, physical servers, etc.) • Third party vendors utilized (if applicable) • Compliance with 21 CFR Part 11 • Data Protection Policy • Change control • Maintenance, backups, business continuity plan • Disaster recovery process & evidence of recent testing	
Day 1/ dd mmm yyyy	12:00pm-1:00pm	Lunch	
Day 1/ dd mmm yyyy	1:00pm-2:30pm	Handling and Processing of Individual Case Safety Reports, including:	

Day#/ Dates	Time- US Eastern Time	Session Description:	Attendees (Name/Title for each session)
		• AE/PQC Training • Sources of information • Translations • Receipt, tracking, duplicate search, source data verification • Handling of follow-up information • Quality Checks • Maintenance of case files and source data • Reconciliation with all sources • Compliance and KPI monitoring *Auditor will select AE/PQC cases to review*	
Day 1/ dd mmm yyyy	2:30pm-4:00pm	Document Review	Auditor
Day 1/ dd mmm yyyy	4:00pm-4:30pm	Review of outstanding requests & Day 1 debrief (if needed)	
		End of Day 1	

Day#/ Dates	Time- US Eastern Time	Session Description:	Attendees (Name/Title for each session)
Day 2/ dd mmm yyyy	8:30am-8:45am	Follow Up from day 1	
Day 2/ dd mmm yyyy	8:45am-9:30am	Project Management Activities • Dedicated Program Manager (if applicable) • Contract Management • Project Meetings • Reporting of key metrics to [sponsor]	

Day#/ Dates	Time- US Eastern Time	Session Description:	Attendees (Name/Title for each session)
Day 2/ dd mmm yyyy	9:30am-10:30am	Overview of Program Services • Patient Access Program • HUB Services • Specialty Pharmacy Services	
Day 2/ dd mmm yyyy	10:30am-10:45am	Break	
Day 2/ dd mmm yyyy	10:45am-11:15am	Review of CAPAs from the [prior years audits yyyy] audit	
Day 2/ dd mmm yyyy	11:15am-12:00pm	Ad Hoc interviews and document review as needed	
Day 2/ dd mmm yyyy	12:00pm-1:00pm	Lunch	
Day 2/ dd mmm yyyy	1:00pm-4:00pm	Document Review & Preparation for Audit Close out	
Day 2/ dd mmm yyyy	4:00pm-4:30pm	Audit Close-out meeting	

Distributor Audit Agenda

Day#/ Dates	Session Description	Date/Time (US EST)	Attendees (Name/Title):
Day 1/ dd mmm yyyy	Opening Meeting • Introductions (auditor and auditee) • Review audit objective and scope • Review audit agenda • Audit logistics	9:00-9:15	
Day 1/ dd mmm yyyy	Company Overview • Overview of company, services, and subcontracting *Please, prepare a short presentation*	9:15-9:45	
Day 1/ dd mmm yyyy	Review of observations from the last audit (yyyy)	9:45-10:00	
Day 1/ dd mmm yyyy	Overview of Learning Management System • Review of AE training deck • LMS - Training System • Review of job descriptions, CVs, and Training Files for [sponsor] project personnel (*auditor will select employee records will be reviewed*)	10:00-10:30	

Day#/ Dates	Session Description	Date/Time (US EST)	Attendees (Name/Title):
Day 1/ dd mmm yyyy	Overview of Quality Management Program • Internal audit program • CAPA processes, management of nonconformances and deviations, customer escalation process	10:30-11:00	
Day 1/ dd mmm yyyy	Break	11:00-11:15	
Day 1/ dd mmm yyyy	Identification and Reporting of Adverse Events (AE) and Product Quality Complaints (PQCs)	11:15-11:30	
Day 1/ dd mmm yyyy	Distribution Services • Packing and Shipping Procedures for [product of focus for audit] • Expedited shipping procedure	11:30-12:00	
Day 1/ dd mmm yyyy	Lunch	12:00-12:30	
Day 1/ dd mmm yyyy	Customer Agreements • Eligibility and Management • Contract Sales and chargebacks (processing and reconciliation) • Invoicing to [sponsor]	12:30-1:00	
Day 1/ dd mmm yyyy	Data Management • 867/852 Report Transmission Process	1:00-1:30	

Day#/ Dates	Session Description	Date/Time (US EST)	Attendees (Name/Title):
Day 1/ dd mmm yyyy	Sales Support, Customer Service and Order Management	1:30-2:00	
Day 1/ dd mmm yyyy	Overview of Computer Systems/Backups/Disaster Recovery/Business Continuity • General info (Name) • Location (web based, physical servers, etc.) • Business Continuity/Disaster Recovery Processes (evidence of testing)	2:00-2:30	
Day 1/ dd mmm yyyy	Document Review	2:30-3:30	
Day 1/ dd mmm yyyy	Audit Close Out	3:30-4:00	

REMs (Prelaunch)

Time (SP-Central Time)	Time (Auditor - Eastern Time)	Topic	Participants
8:00-8:30	9:00-9:30	Opening Meeting • Introductions (auditor and auditee) • Review audit objective and scope • Review audit agenda • Audit logistics	
8:30-8:45	9:30-9:45	Company Overview • History and overview (*please provide a brief presentation*) • Services, sponsors, organizational structure, alliances/partnerships, resources • Regulatory inspection history	
8:45-9:30	09:45 – 10:30	Quality Management System • QMS overview • Document Control • Training: New Employee Onboarding, Annual training, training curriculum, LMS, etc. • Job descriptions and Curriculum Vitae (CVs) • Management of audits and inspections • Deviations and CAPA management • Documentation, archiving & record management (paper vs. electronic)	

Time (SP-Central Time)	Time (Auditor - Eastern Time)	Topic	Participants
9:30-10:00	10:30-11:00	Break	
10:00-11:00	11:00-12:00	Overview of the Pharmacovigilance System • Medical Information Inquiries • AE and PQC ICSR handling • QC Process • Reconciliation • Reporting Compliance *Auditor will select AE/PQC cases for review.*	
11:00-12:00	12:00 – 1:00	[add product name] REMS Program • [product name] REMS Authorized Pharmacy Representative and additional staffing • [product name] REMS specific processes and procedures • [product name] REMS Training Program (training materials, review of selected CVs, JDs, and training records) • Adverse Event Reporting Process • Reconciliation Process • [product name] REMS Reporting	

Time (SP-Central Time)	Time (Auditor - Eastern Time)	Topic	Participants
12:00-1:00	1:00-2:00	Lunch	
1:00-2:00	2:00-3:00	[product name] REMS continued: • Monitoring of REMS program compliance • Management of REMs noncompliance • Communication with [sponsor] • Process for change to the Pharmacy Authorized Representative • Additional processes and procedures (if applicable)	
2:00-3:30	3:00-4:30	Document Review	Auditor
3:30-4:00	4:30-5:00	Day 1 debrief and review of outstanding requests (if applicable)	

Audit Day 2: MMDDDYYYY

Time (SP-Central Time)	Time (Auditor - Eastern Time)	Topic	Participants
8:00-8:30	9:00 – 9:30	Review of outstanding requests from Day 1	
9:00-10:00	10:00 – 11:00	Overview of Computer Systems • General info (Names, versions, etc.) • Location (web based, physical servers, etc.) • Third party vendors utilized (if applicable) • Validation (21 CFR Part 11) • Data Protection Policy • System Security, Maintenance, backups • Change control ○ Business continuity plan ○ Disaster recovery process & evidence of recent testing	
10:00-10:15	11:00-11:15	Break	
10:15-12:00	11:15-1:00	REMS discussion continued/document review	
12:00-1:00	1:00 – 2:00	Lunch	
1:00-3:30	2:00 – 4:30	Document Review and Preparation for Closing Meeting	
3:30-4:00	4:30 – 5:00	Closing Meeting	

REMS (Postlaunch/Maintenance)

AUDITEE:

TYPE OF AUDIT:

AUDIT PURPOSE:

AUDIT DATE(S):

AUDIT STANDARDS:

AUDITOR(S):

Audit Day 1

Time EST	Time CST	Topic	Participants
8:00-8:30	9:00-9:30	Opening Meeting • Introductions (auditor and auditee) • Review audit objective and scope • Review audit agenda • Audit logistics	
8:30-8:45	9:30-9:45	Company Overview • History and overview (*please provide a brief presentation*) • Services, sponsors, organizational structure, alliances/partnerships, resources • Subcontracting (third party suppliers) • Regulatory inspection history	
8:45-9:30	09:45 – 10:30	Quality Management System • QMS overview • Document Control • Training: New Employee Onboarding, Annual training, training curriculum, LMS, etc. • Job descriptions and Curriculum Vitae (CVs) • Management of audits and inspections • Deviations and CAPA management • Documentation, archiving & record management (paper vs. electronic)	

Time EST	Time CST	Topic	Participants
9:30-10:00	10:30-11:00	Break	
10:00-11:00	11:00-12:00	Overview of the Pharmacovigilance System • Medical Information Inquiries • AE and PQC ICSR handling • QC Process • Reconciliation • Reporting Compliance *Auditor will select AE/PQC cases for review.*	
11:00-12:00	12:00 – 1:00	[add product name] REMS Program • XX …REMS Authorized Pharmacy Representative and additional staffing • XX….REMS specific processes and procedures • REMS Training Program (training materials, review of selected CVs, JDs, and training records) • Adverse Event Reporting Process • Reconciliation Process • XX….REMS Reporting	
12:00-1:00	1:00-2:00	Lunch	
1:00-2:00	2:00-3:00	XXXX….. REMS continued: • Monitoring of REMS program compliance • Management of REMs noncompliance • Communication with client company • Process for change to the Pharmacy Authorized Representative • Additional processes and procedures (if applicable)	
2:00-3:30	3:00-4:30	Document Review	

Time EST	Time CST	Topic	Participants
3:30-4:00	4:30-5:00	Day 1 debrief and review of outstanding requests (if applicable)	

Audit Day 2

Time	Time	Topic	Participants
8:00-8:30	9:00 – 9:30	Review of outstanding requests from Day 1	
8:30-9:00	9:30-10:00	Review of the Dispensing Process for [product] • [product] REMS Dispensing Program/System Demo	
9:00-10:00	10:00 – 11:00	Overview of Computer Systems • General info (Names, versions, etc.) • Location (web based, physical servers, etc.) • Third party vendors utilized (if applicable) • Validation (21 CFR Part 11) • Data Protection Policy • System Security, Maintenance, backups • Change control o Business continuity plan o Disaster recovery process & evidence of recent testing	

Time	Time	Topic	Participants
10:00-10:15	11:00-11:15	Break	
10:15-12:00	11:15-1:00	REMS discussion continued/document review	
12:00-1:00	1:00 – 2:00	Lunch	
1:00-3:30	2:00 – 4:30	Document Review and Preparation for Closing Meeting	
3:30-4:00	4:30 – 5:00	Closing Meeting	

Preaudit Document Request List Templates

Preaudit Document Request List: Example1

Initial Document requests: Please provide by DDMMMYYYY

- Pharmacovigilance (PV) system Organizational Charts showing interactions with other functions.
- Inspection history.
- List of third-party service providers. Please include company name, service, territory, start date, end date, short description (e.g., responsibilities/activities covered).
- List(s) of PV-related computerized systems. Please include system name, version, validation status, production (go-live) date, last upgrade date, and a description of the system.
- List of QMS and PV system key performance indicators (KPIs).
- List(s) of Procedures: quality documents (QDs), operating procedures, working instructions, policies, manuals, guidance and other relevant tools and training items related to the PV, IT and QMS activities. Please include ID, title/name, version#, effective date. *Auditor will then select procedures to be reviewed prior to the audit.*
- List(s) of [sponsor] Project PV-related deviations/CAPAs and their associated action items (corrections/corrective actions/preventive actions/action items) associated with the [sponsor]project. Please include ID, title, short description, date of awareness, date of reporting/opening, date of closure, type (e.g., deviations, CAPA plans, effectiveness checks, escalated issues, change controls) and categorization.
- List of [vendor] personnel involved in the [sponsor] programs. Please list per program with job title and start and end dates on the [sponsor] programs. *Auditor will then select JDs, CVs, and training records to be reviewed for personnel.*

- Training matrices and training compliance for all roles performing PV activities. Please include total # of Trainings completed "on-time", total # of Trainings Due in the period (e.g., Month), On-Time Training Compliance %.
- List of all documents specific to the [sponsor] program.
- List of all computer systems utilized for [sponsor] activities.
- Line listing of AEs/PQCs for the [sponsor program] (DDDMMMYYY–DDMMMYYYY).

Preaudit Document Request List: Example 2

No	Description or Name of Document(s) Requested:	Comments
1	Organizational chart for [vendor]	
2	Full List of [vendor] SOPs,WIs, Policies	
3	[vendor] procedures for: • Quality manual (if applicable) • Controlled Document management • Training Program • Deviation and CAPA Management • Internal audit program • Vendor selection and management • Record retention • AE/PQC Corporate Policy • SOP for AE/PC Handling and Management • IT System Security • IT Change Control • IT System monitoring and maintenance • Business Continuity • Disaster Recovery & evidence of most recent testing *Please note that upon receipt the auditor may request additional documents*	
3	List of all [vendor] personnel (with job title) for the [sponsor] project *(Once received, the auditor will select those personnel to review JDs, CVs, and training records)*	

No	Description or Name of Document(s) Requested:	Comments
4	List of all contracts and PV agreements with [sponsor] that would be in scope for this audit.	
5	Listing (Excel) of Individual Case Safety Reports from DDMMMYYYY to DDMMMYYYY. *(Once received the auditor will select cases for review in your system during the audit)*	
6	Listing (Excel) of Product Quality Complaints (PQCs) received from DDMMMYYYY to DDMMMYYYY. *(Once received the auditor will select cases for review in your system during the audit)*	
7	[sponsor project] Compliance metrics reports from DDMMMYYYY to DDMMMYYYY (if applicable).	
8	List of [sponsor project] Deviations/CAPAs from DDMMMYYYY to DDMMMYYYY. Please indicate whether each is open or closed.	
9	List of computer systems utilized in support of the [sponsor] project activities including: ○ Name and version of the system ○ Validation status as per FDA 21CFR Part 11/EMA Annex 11 ○ Date of validation ○ Date of revalidation (if applicable) ○ Validation Summary Reports for each system	
10	Validation Summary Reports for computer systems utilized in #9 above.	

No	Description or Name of Document(s) Requested:	Comments
11	AE/PQC Reconciliation (add number of reports) for: Month YYYY Month YYYY Month YYYY Month YYYY Month YYYY	

AE/PQC Reconciliation Templates

AE/PQC Reconciliation Template: Example 1

AE/PQC Reconciliation Summary											
Case Number	Initial Awareness Date	Report Awareness Date	Case Priority	Destination	Report Form	Follow Up #	Product	License Number	Due Date	Submission Date	Reconciled (Yes or No)

AE/PQC Reconciliation Template: Example 2

Vendor Case ID	Vendor Patient ID	Sponsor Case ID	Patient Initials	DOB	Gender	Drug Name	Date AE Identified	AE/PQC Description	Date sent to Sponsor	Reconciled (Yes or No)

Audit Report Templates

Full Narrative Report

Audit Date(s):	*DDMMMYYYY to DDMMMYYYY*
Auditee:	*Name, address and primary contact information*
Audit Location (s):	Onsite add address otherwise indicate if remote
Audit Type:	☐ Pre-qualification/Qualification Audit ☐ Re-qualification/Routine Audit ☐ Internal Audit ☐ External Audit ☐ Third Party ☐ For-cause/Directed Audit
Auditor (s)/SME:	Add lead auditor name and address
Audit Report Distribution:	Add names with job titles
Signature:	

Introduction

A [*routine/for cause onsite/ remote]* Pharmacovigilance audit of [*auditee*] was conducted on *DDMMMYYYY to DDMMMYYYY*. The audit assessed the compliance and adequacy of [*auditee*] activities specific to [*list primary contracted activities*] with regards to Pharmacovigilance (PV) regulatory requirements, contractual obligations, industry best practices, project procedures and [*auditee*] Standard Operating Procedures (SOPs) and Work Instructions as applicable.

Objective

The objective of this audit was to ensure [*auditee*] had the appropriate capabilities to comply with regulations and guidelines, relevant procedural documents, and contractual requirements.

Executive Summary

A [*onsite/remote*] Pharmacovigilance (PV) audit of [auditee] was conducted DDMMMYYYY to DDMMMYYYY by [*add auditor name and job title*]
The [*auditee*] audit included a review of the following: *based on agenda*

Quality Management System (QMS)
Individual Case Safety Report (ICSR) Management and Submissions
Literature Search Activities
Aggregate Report Management
Safety Governance
Computer systems/applications
Business Continuity and Disaster Recovery

Based on the scope of the audit, documentation and evidence provided as well as interviews with subject matter experts (SMEs), no [*or add the number of Critical, Major, or Minor*] observations were identified.

List Critical and Major observations and provide a brief description:

Background

Provide a brief background on the auditee company based on the company overview session conducted during the audit.

Opening Meeting

An opening meeting was led by the audit host, [*add name with job title*] on *DDMMMYYYY*. The following items were reviewed - Audit objectives and scope, review of audit agenda, and audit logistics.

Documentation Review

The following documents were reviewed:

Provide in the following format:

Document Number
Document Name
Version
Effective date

Personnel Interviews

The following [auditee] individuals participated in the audit:

Name with job title

Closing Meeting

A closing meeting was held on *DDMMMYYYY* during which potential observations were reviewed and the process for the provision of the audit report and CAPA report were reviewed.

SUMMARY OF AUDIT SYSTEMS REVIEWED

The following sections contain summaries of the primary systems reviewed during the audit. These summaries illustrated a high-level discussion of the systems/controls in place to support [*sponsor*] contracted activities and were not intended to provide a detailed description of all documents and evidence reviewed as well as interview discussions conducted.

Quality Management System

Controlled Document Management

Provide an overview of document and evidence reviewed describing how controlled documents are managed by the auditee. Indicate whether there are any areas of noncompliance resulting in observations.

Training

Provide an overview of documents and evidence reviewed describing how this topic is managed by the auditee. Indicate whether there are any areas of noncompliance resulting in observations.

Job Descriptions, Curriculum Vitaes and Training Records for the following [*auditee*] project team members were reviewed:

Name with job titles

Deviations/Corrective and Preventative Action (CAPAs)

Provide an overview of documents and evidence reviewed describing how this topic is managed by the auditee. Indicate whether there are any areas of noncompliance resulting in observations.

Records Retention

Provide an overview of document and evidence reviewed describing how this topic is managed by the auditee. Indicate whether there are any areas of noncompliance resulting in observations.

Adverse Event (AE)/Product Quality Complaint (PQC) Management

Provide an overview of documents and evidence reviewed describing how this topic is managed by the auditee. Indicate whether there are any areas of noncompliance resulting in observations.

List AE/PQC cases reviewed:

SAE Reconciliation

Provide an overview of documents and evidence reviewed describing how this topic is managed by the auditee. Indicate whether there are any areas of noncompliance resulting in observations.

List SAE reports reviewed:

Literature Search and Analysis

Provide an overview of documents and evidence reviewed describing how this topic is managed by the auditee. Indicate whether there are any areas of noncompliance resulting in observations.

Aggregate Reports

Provide an overview of documents and evidence reviewed describing how this topic is managed by the auditee. Indicate whether there are any areas of noncompliance resulting in observations.

Benefit Risk Management

Provide an overview of documents and evidence reviewed describing how this topic is managed by the auditee. Indicate whether there are any areas of noncompliance resulting in observations.

Computer Systems and Applications

Provide an overview of documents and evidence reviewed describing how this topic is managed by the auditee. Indicate whether there are any areas of noncompliance resulting in observations.

List each computer system used to support [sponsor] activities and provide an overview of each system

Business Continuity Planning (BCP) and Disaster Recovery (DR)

Provide an overview of documents and evidence reviewed describing how this topic is managed by the auditee. Indicate whether there are any areas of noncompliance resulting in observations.

Audit Observations

Complete the following for each observation:

Observation Category
Observation Sub Category
Observation Description
References

Abbreviated Report

Audit Dates: DDMMMYYYY-DDMMMYYYY

Auditor: Name with job title

Auditee Name and Address:

Main Auditee Contact: Name, job title, email address

Type of Audit: (remote or onsite)

Executive Summary

This was a routine audit of pharmacovigilance activities performed by the [*auditee*] and [*sponsor*] to assess the compliance of the pharmacovigilance system with applicable global and local regulations, and to review the interfaces and interdependencies between [*sponsor*] and the [auditee].

The scope of the audit covered the services outlined in the Master Services Agreement (MSA) and associated Statement of Work (s) (SOWs).

The results of the audit revealed (#) critical observations, (#) major observations and (#) minor observations.

The critical/ major observations were : (*provide a brief description*)

Conclusions: *Summarize the auditees performance with identified issues for remediation*

Audit Conduct

Introduction:

This *routine/for cause* PV audit of [*add auditee name*] was remotely (*if onsite add address of location*) conducted on DDMMMYYYY-DDMMMYYYY to assure that the PV activities conducted by [*auditee*] relating to the management of [*add services*] as outlined in the Master Services Agreement (MSA) and associated Statement of Work(s) [*add additional documents*] were reviewed.

The audit team included:

Name, job title of auditor and other audit team members

Data/Documents Reviewed:

List in the following format

Document ID Number

Document Name

Version

Effective date

Observations:

Create one table for each observation. Each observation to include the following:

Category

Sub-Category

Risk Classification
Title of the Observation (brief statement)
Issue (describe the issue with objective evidence)
References: List references specific to support the observation

Annual Audit Report

Audit Number

Type of Audit Routine/For Cause

Audit Format Remote/Onsite

Planned/Unplanned

Institution/Site Name

Site Address

Main Contact

Audit Dates DDMMMYYYY to DDMMMYYYY

Scope: The audit scope included the….. *list as per the audit agenda.*

Purpose: *Describe based on audit scope. For example, "The purpose of this audit was to evaluate compliance with the applicable procedural documents and regulations as well as evaluate the effectiveness of previous corrective and preventive actions following previous audit and determine if [sponsor name] activities were being performed as contracted".*

AUDIT CONDUCT

Executive Summary

[*vendor name*] was contracted to provide [*add services*] for [*add product in scope*] in the USA.

The primary objectives of this audit were to: *List in bullet form*

Upon completion of the audit there were (#) Critical, (#) Major, (#) Minor Observations and (#) recommendations.

Provide brief description of each observation….

<u>Main Body of the Report</u>

Audit Participants:

This remote audit was conducted by [*add auditor name*] on DDMMMYYYY to DDMMMYYYY via [*add platform name if remote otherwise add location for onsite audit*].

The following [sponsor] employees were involved during the conduct of the audit:
List with name and job title

Opening Meeting:

The opening meeting was attended by the Audit Host [*names and job titles*]

All audit requests were provided via [*add platform or method*].

Company Updates since the Last (YYYY) Audit:
List in bullet form

QUALITY MANAGEMENT SYSTEM: UPDATES SINCE THE LAST (YYYY) AUDIT
Specify each section with narrative including changes

ADVERSE EVENT (AE)/PRODUCT QUALITY COMPLAINT (PQC) MANAGEMENT UPDATES SINCE THE LAST (YYYY) AUDIT

Specify each section with narrative including changes
AE/PQC Cases reviewed (List case numbers)
List reconciliation reports reviewed

COMPUTER SYSTEMS / DATA SECURITY / BACKUPS FOR [sponsor] UPDATES SINCE THE LAST (YYYY) AUDIT:

Specify each section with narrative including changes

BUSINESS CONTINUITY PLANNING (BCP)/ DISASTER RECOVERY (DR):

Specify each section with narrative including changes

DOCUMENTS REVIEWED:

List in the following format : Document ID #, Document Name, Version, Effective date

EXIT MEETING:

Exit meeting participants included the following [vendor/sponsor] personnel:

List with names and job titles

CONCLUSIONS:

Provide conclusions of the audit including overall performance of the auditee and summarize findings.

OBSERVATIONS:

Description of Observation: Describe the observation and document objective evidence.

Observation Severity: Critical, Major, or Minor

Observation Category:

Observation Sub-Category:

References:

Repeat information for each observation.

CAPA Plan Templates

CAPA Plan Template: Example 1

Corrective Action Preventive Action (CAPA) Plan

<table>
<tr><td colspan="3">Observation 1 – add Category</td></tr>
<tr><td colspan="3">Observation Summary: Provide a summary statement of the observation

Observation Description: Provide details of the observation including supportive objective evidence

References: List references applicable to the observation</td></tr>
<tr><td colspan="3">Root Cause: Auditee to provide a root cause analysis of the observation. There can be one than one root cause.</td></tr>
<tr><td>Corrective Action(s) and Preventative Action(s)</td><td>Deliverable(s) Document deliverables/evidence to address the corrective and preventative actions with associated dates for completion</td><td>Responsible person (s) &
Due Date(s)</td></tr>
<tr><td></td><td></td><td></td></tr>
<tr><td></td><td></td><td></td></tr>
</table>

Observation 2 – *add Category*		
Observation Summary: Provide a summary statement of the observation *Observation Description: Provide details of the observation including supportive objective evidence* *References: List references applicable to the observation*		
Root Cause*: Auditee to provide a root cause analysis of the observation. There can be one than one root cause.*		
Corrective Action(s) and Preventative Action(s)	Deliverable(s) *Document deliverables/evidence to address the corrective and preventative actions with associated dates for completion*	Responsible person (s) & Due Date(s)

CAPA Plan Template: Example 2

Audit: ______________________________

Audit Date: ______________________________

CAPA Plan Due Date: ______________________

Observation Number	Observation Description	Root Cause (Systemic or Isolated as applicable)	Corrective Action	Preventive Action	Responsible Person	Completion Date (Actual or Anticipated)
1.0	*Add Category of Observation*					
1.1	Add observation including objective evidence provided to confirm observation					
1.2						
2.0	Quality Management System					
2.1	Add observation including objective evidence provided to confirm observation					
3.0	Records Management					
3.1						

Index

H. Murphy, *A Guide to Pharmacovigilance Audits in the USA*, https://doi.org/10.1007/978-3-031-73596-7

The manufacturer's authorised representative in the EU is Springer Nature Customer Service Centre GmbH, Europaplatz 3, 69115 Heidelberg, Germany. If you have any concerns regarding our products, please contact ProductSafety@springernature.com

Printed and bound by CPI Group (UK) Ltd, Croydon, CR0 4YY
01/10/2026
02225875-0008